T0181698

Springer Series on Touch and Haptic Systems

More information about this series at http://www.springer.com/series/8786

Femke Elise van Beek

Making Sense of Haptics

Fundamentals of Perception and Implications
for Device Design

 Springer

Femke Elise van Beek
Vrije Universiteit Amsterdam
Amsterdam, The Netherlands

ISSN 2192-2977   ISSN 2192-2985   (electronic)
Springer Series on Touch and Haptic Systems
ISBN 978-3-319-88863-7   ISBN 978-3-319-69920-2   (eBook)
https://doi.org/10.1007/978-3-319-69920-2

Printed on acid-free paper

This Springer imprint is published by Springer Nature
The registered company is Springer International Publishing AG
The registered company address is: Gewerbestrasse 11, 6330 Cham, Switzerland

This book is dedicated to my father, Jörg van Beek. He made the beautiful cover for my thesis, upon which this book is based.

Dikke kus papa.

Acknowledgements

There are so many people that I should be thanking here for helping me along my path to obtaining a PhD degree. Luckily, I thanked most of them already in my thesis acknowledgements. However, you cannot thank amazing promoters and supervisors often enough, so Astrid Kappers and Wouter Bergmann Tiest, thank you again! Without you, this book certainly would not have existed.

The research in my thesis, upon which this book is based, was performed at the Physics of Man Group of Utrecht University, Helmholtz Institute, and at the Faculty of Behavioural and Movement Sciences of the Vrije Universiteit Amsterdam, MOVE Research Institute Amsterdam, the Netherlands. This research was part of the H-Haptics programme and was supported by the Dutch Technology Foundation STW, which is part of the Netherlands Organisation for Scientific Research (NWO) and partly funded by the Ministry of Economic Affairs (project number 12157).

Contents

Chapter 1
Introduction

Abstract There are many environments in which it is not practical to do things with your own hands directly. For instance, it is not wise to enter a nuclear plant to perform device maintenance. In that case, it would be smarter to position robots inside the plant that can be directed by the maintenance worker outside of the plant. Another example is a keyhole procedure in surgery. In this procedure, instruments are inserted into a patient through tiny holes, to avoid large wounds in relatively simple procedures. The surgeon could use his own hands to guide the tools, but quite often, the surgeon actually uses a joystick to guide a robot to insert the tools, because this enables him to scale his movements to small and precise movements inside the patient. So, both the maintenance worker and the surgeon use teleoperation systems: systems consisting of a *master*, which is the interface (such as a joystick) that is used by a human (from now on called *operator*), and a *slave*, which is the robot that is performing the action in the remote environment (Srinivasan and Basdogan, Comput Graph 21(4):393–404, 1997). Obviously, there are large advantages to teleoperation techniques for the operators: the maintenance worker in the nuclear plant is not exposed to radiation and the surgeon has a far better view of his patient, because his hands are not in the way. However, there is also at least one large disadvantage to this technique: since the master and the slave device are usually not directly connected, the operator cannot directly feel what the slave is doing. When he would be performing the task with his own hands, he could have used sensory information from his hands, also called *haptic perception*, to feel what the slave is doing. To solve this problem, haptic feedback can be incorporated in the master device, which is the virtual equivalent of natural haptic information.

The aim of this book is to investigate parameters in haptic perception that are important for designing haptic devices and haptic feedback. This is done by using a deductive approach: fundamental properties of haptic perception are investigated, which can be applied to the design of specific haptic devices later on [9]. Haptic device designers usually take the inductive approach: by testing the performance of their specific devices in user studies, they obtain general guidelines on which device parameters ensure satisfactory human performance [7]. However, when

© Springer International Publishing AG 2017

F.E. van Beek, *Making Sense of Haptics*, Springer Series on Touch and Haptic Systems, https://doi.org/10.1007/978-3-319-69920-2_1

doing this, it is hard to understand why certain approaches work and others do not, which limits the generalizability of the findings. Therefore, this book intends to provide the fundamental knowledge that is needed in order to solve these questions from a deductive perspective. This is done by answering questions like: if humans move their hands, how precisely can they perceive the distance that they have covered? The answers to these questions provide fundamental knowledge on haptic perception, which can be used in the design of a whole range of devices.

1.1 Teleoperation

When performing precise operations, there are two important sources of information: what you see with your eyes and feel with your hands, or in other words, visual and haptic information. When the operator and the slave are separated, the operator cannot directly acquire this information any more, so this information needs to be provided in another way. In the current teleoperation systems, there is often a pretty good representation of visual information from the slave side, using multiple cameras and even 3D images, which together provide a visual experience which sometimes even outperforms the information that the operator would normally receive when looking at the environment directly. However, designing haptic feedback in such a way that it resembles natural haptic feedback is still a big challenge. Because of technical limitations, it is still impossible to build haptic devices that can provide operators with haptic feedback which is similar to the feedback that they would have received when they had been performing the task with their own hands [6, 20]. Nonetheless, it might not be necessary to re-create all the natural feedback, because the haptic sense also has limitations. So, part of the answer to the challenge of designing applications more efficiently could be to take human perception into account [5, 21, 26].

Apart from designing haptic feedback which resembles natural feedback as closely as possible, force feedback could also be used to provide the operator with extra information or to guide the operator towards a target position by adding extra forces. These are the concepts of haptic guidance or haptic shared control, in which the goal is that the human and machine perform a task together [1]. Usually, guidance forces are presented as an attractive force field around a target or as a tunnel which helps operators to stay on a desired trajectory. Haptic guidance mostly improves performance in terms of parameters like task completion time [15]. However, studies also often report conflicts between human and machine when guidance forces are designed in this relatively simple way (e.g. [3, 14]). Especially for tasks involving motor learning, it has been reported that using haptic guidance can even deteriorate task performance [18]. Apparently, designing guidance forces in this rather simple way is not always the optimal solution for the human user. Again, information on human perception could provide an answer to this. In the next section, a general introduction in haptic perception will be given.

1.2 Haptic Perception

Haptic perception actually covers two perceptual subsystems: cutaneous perception, which refers to the sense of touch, and kinesthetic perception, which refers to the sense of body position and movement [12, 13]. In cutaneous perception information is provided by two types of receptors, embedded in the skin: thermoreceptors and mechanoreceptors. Thermoreceptors are sensitive to temperature, while mechanoreceptors are sensitive to deformation caused by force or displacement. There are two types of thermoreceptors in the skin, which respond to warmth and cold. There are four types of mechanoreceptors in the skin, which all contribute to cutaneous perception: Meissner corpuscules, Merkel cell complexes, Ruffini endings and Pacini corpuscules. These four types of receptors can be categorized according to two properties: their receptive field size ('small' and 'large', also referred to as 'type I' and 'type II') and their adaptation rate ('slow' and 'fast', also referred to as 'SA' and 'FA') [10]. The receptive field size refers to the size of the skin surface from which a receptor obtains information, while the adaptation rate indicates how fast a receptor becomes insensitive to a stimulus when the stimulus does not change. By combining these different sensory properties, information about temporal and spatial characteristics of the touched object can be derived. For instance, fast adapting sensors are useful for detecting changes in the touched object, while the slow adapting types provide information about more stationary properties of the object.

Within the muscles and joints, there are three types of mechanoreceptors, which are the main receptors responsible for providing kinesthetic information: muscle spindles, Golgi tendon organs and joint receptors [17]. Joint receptors provide information about joint angles based on the stretch and strain of the tissue inside the joints, while muscle spindles and Golgi tendon organs provide information about the state of the muscle and the resulting state of the tendon, from which arm position and movement can be inferred. Together, these receptors provide an image of body posture and movement.

Haptic perception can be investigated at many levels, ranging from recordings inside single neurons to experiments at a behavioural level. The approach in this work is a psychophysical one: by investigating the relation between the physical properties of a stimulus and the perception by a participant of that stimulus, regularities in these relations can be discovered, which allows for a general description of the relation between stimulus properties and human perception [8].

1.3 Psychophysics

In psychophysical research, the aim is to establish a relation between the properties of a stimulus and the perceptual experience of that stimulus [8]. If this relation is known, predictions can be made about other stimuli than the tested ones, and these predictions can be tested again to validate the model. In psychophysical experiments, humans are usually asked to rate the property of a stimulus, either by rating the stimulus on its own (i.e. 'how heavy is this cube?') or by comparing two stimuli (i.e. 'which of the two cubes is heavier?'). When using the former method, the main property that can be derived is stimulus intensity. A very common procedure for these types of experiments is free magnitude estimation. For instance, when cubes with different weights are used, participants are asked to rate the heaviness of each of the cubes. Usually, they are free to chose their own scale for this rating. These types of experiments can reveal the relationship between physical and perceived stimulus intensities.

When using the method of comparing two stimuli, usually one stimulus is the reference, which is kept constant throughout the experiment, while the other is the test, which varies in order to investigate the results of the variations of the property. The difference between the test and the reference stimulus determines the ease with which the participant can differentiate between the two stimuli. In the example with the two cubes, it is easy to imagine that if the reference is much lighter than the test, the participant will always answer that the test is the heavier one. If the reference is much heavier than the test, the participant will always answer that the reference is the heavier one. In between those extremes, there is a gradual transition from one answer to the other, which is called the psychometric curve. When assuming that the answers follow a normal Gaussian distribution, the psychometric curve can be described using a cumulative Gaussian distribution [8]. An example of such a curve is shown in Fig. 1.1.

From this curve, important perceptual properties can be inferred, which are: the Point of Subjective Equality (PSE, see Fig. 1.1) and the discrimination threshold ('DT' in Fig. 1.1). The PSE refers to the intensity of the test stimulus at which it is perceptually equal to the reference stimulus, which is the point where the response fraction is 50%. At this point, both stimuli feel equally heavy and thus the participant must guess. The bias ('B' in Fig. 1.1) is the difference between the PSE and the reference stimulus intensity. The larger the bias, the lower the perceptual accuracy.

When using a reference and a test cube that only differ in weight, while all the other properties are exactly the same, no bias is expected. However, when a property other than weight differs between the test and the reference stimulus, the effect of this property on weight perception can be assessed by looking at the bias. An example of a situation in which a bias can be expected, is when the reference cube is composed of a different material than the test cube and, as a consequence, has a larger volume, while weighing the same. In this situation, the reference cube is probably perceived as being lighter than the test cube because of the size-weight illusion [11]. The size of the illusory effect can thus be inferred from the size of the bias. Biases between different senses are also frequently observed, such as the visuo-haptic bias: when moving your unseen hand to a visual target, you usually do not end up at the physical location of the visual target, so there is a difference between the physical and the perceived target location [19]. In Chap. 8, such visuo-haptic biases are investigated.

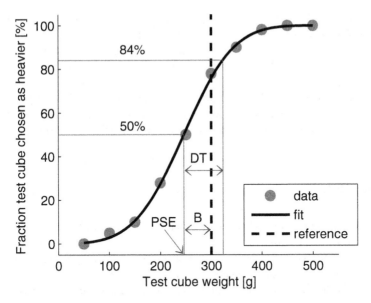

Fig. 1.1 Example of a psychometric curve. For this hypothetical experiment, participants are asked to judge which of two presented cubes is heavier. On each trial, a reference stimulus with a constant weight (dashed line) and a test stimulus with varying weight (depicted on the horizontal axis) is presented. The reference stimulus is composed of another material than the test stimulus, which affects its perceived weight. For each test stimulus intensity, the fraction with which this stimulus is perceived as the heavier one is shown (gray dots). By fitting a psychometric curve to these data (thick black line), the PSE and the discrimination threshold (DT) can be inferred. The bias (B), which is the difference between the PSE and the reference stimulus, represents the perceptual accuracy, while the discrimination threshold represents the perceptual precision

The discrimination threshold describes the difference in stimulus intensity that is needed to reliably determine that the stimuli are different. In this book, a response fraction of 84% is used as the discrimination threshold. When a bias is present, the discrimination threshold is defined as the difference between the test stimulus intensity corresponding to a 84% response fraction and the Point of Subjective Equality, as can been seen in Fig. 1.1. For higher discrimination thresholds, the perceptual precision is lower and the psychometric curve is flatter. The discrimination threshold is related to the perceptual noise. If our receptors and perceptual processing were noiseless, the discrimination threshold would be very small, since tiny differences in stimulus intensity would be noticeable in that situation. This is usually not true, so therefore, the discrimination threshold is an interesting parameter. The discrimination threshold is also commonly expressed as a fraction of the reference stimulus intensity, which is called the Weber fraction. Weber's law states that the ratio between the discrimination threshold and the reference stimulus intensity is constant [27]. If the ratio between the two stays constant, this means that the absolute discrimination threshold increases with stimulus intensity. Although this law has proven to be wrong for some stimulus properties (such as the perception of symmetry [25]), it generally holds for most medium-sized stimulus intensities. However, for small stimulus intensities a floor effect on the absolute discrimination thresholds is often observed, leading to an increase in Weber fractions for these small intensities [4, 22, 24]. The discrimination threshold can be influenced in many ways. For instance, when the perception is restricted to be passive, which means that the stimulus is applied to the participant's hand, without him/her performing any active movements, the discrimination threshold is usually higher than when the participant is allowed to move freely and thus to perceive the stimulus actively [23].

1.4   Outline

In this book, the precision and accuracy of haptic perception of several properties is investigated by determining biases and discrimination thresholds in psychophysical experiments. This work can be divided into three parts, which together cover different aspects of haptic perception and move from fundamental to more applied topics. The topics of the parts are: static haptic perception, dynamic haptic perception, and implications of haptic biases for haptic devices.

The first part consists of three chapters, Chaps. 2, 3, and 4, which are all concerned with haptic perception under static circumstances, so stimuli are applied to the stationary hand of a participant. A logical question in this context is: how does a participant perceive that someone is pulling or pushing his/her hand? In all three chapters, the task for the participant was to perceive forces which were exerted on his/her hand, while (s)he had to keep his/her hand in the same position. In Chap. 2, a study on the effect of force direction on force perception in 2D is presented. In this chapter, biases in the perception of force direction and force magnitude were studied. These experiments show direction-dependent biases in force magnitude perception which were consistent across participants, while biases in force direction perception were also direction-dependent, but much more variable across participants. To further investigate the variable biases in force direction perception, the study in Chap. 3 was designed, in which the nature and consistency of these direction-dependent patterns were examined. By studying a small group of participants at consecutive moments in time, the consistency of the biases within participants over time was investigated. By studying a large group of participants during one session, the consistency of the patterns across participants was investigated. Chapter 4 describes another follow-up study on Chap. 2, in which the investigation of biases in force magnitude perception was extended to a 3D situation. Moreover, since the patterns were very consistent across participants, a hypothesis to explain the nature of the biases was tested. This hypothesis was based on the notion that biomechanical parameters of the arm also show a direction-dependency, which is caused by the anatomy of the muscles, tendons, bones, and other tissues. The direction-dependency of biomechanical parameters of the arm seemed to align fairly well with the direction-dependency of the perceptual biases found in Chap. 2. To test this hypothesis, these biomechanical parameters of the arm were measured in Chap. 4 using system identification techniques. By also measuring the perceptual biases in force magnitude perception in this study, they could be compared directly to the measured biomechanical parameters to test the hypothesis that the latter were responsible for the biases in force magnitude perception.

The second part consists of two chapters, Chaps. 5 and 6, which describe parameters that become important in dynamical situations, so when humans start to move their arm. Important aspects of movement are: the perception of one's own movements and the interaction with objects using movement. Chapter 5 is focussed on the perception of arm movement, by studying discrimination thresholds for movement distance for various types of arm movements. The research in this chapter tests if discrimination thresholds are affected by movement distance, movement direction, and stimulus type. Furthermore, a passive condition, in which a haptic device moved the participant's arm, was compared to an active condition, in which the device was moved by the participant. Chapter 6 revolves around the interaction with objects when moving. In particular, the perception of object hardness was investigated. It is already known that stiffness is very important in hardness perception [2], but this chapter focusses on another object property, which is damping. This is also interesting for teleoperation applications. Teleoperation systems usually involve a delay between sending information from master to slave and back again, which can cause instabilities in the system. To avoid this, damping is often injected in delayed teleoperation systems. In this chapter, the effect of adding damping on the biases in the operator's perception of the hardness of objects within that system is investigated. Several levels of damping and object stiffness were combined in order to be able to construct an overview of the relation between stiffness, damping and perceived hardness.

The third part consists of two chapters, Chaps. 7 and 8, which are more directly related to designing haptic guidance for teleoperation applications. Nonetheless, they also answer questions that are interesting from a fundamental perspective, since they provide knowledge on the integration of sensory information. In Chap. 7, the integration of position and force information during the perception of force fields is described. Both the biases and the variability (which is related to the discrimination threshold) of the data were investigated. Two hypotheses were tested, which were both described in a mathematical model and thus could both be used to predict biases and variability. By comparing the experimental data with the model predictions, the validity of the hypotheses could be tested. In Chap. 8, the well-known paradigm of visuo-haptic biases [19] was used to test the use of correcting for user-specific perceptual biases in the design of haptic guidance. It has already been shown that correcting the mapping between operator and slave movements, by using parameters that are consistent across participants, increases user performance [16]. However, the sizes of perceptual biases often differ between participants, and several types of biases are even completely user-specific (such as the biases in perception of force direction, which are described in the first part). The study in this chapter describes a comparison between a task in which physically correct haptic guidance was presented and one in which the haptic guidance was adjusted to the user-specific biases. Both the biases and the variability of the data were investigated.

Together, the three parts provide knowledge on fundamental parameters of haptic perception that can be important in the design of haptic devices. In the General Discussion chapter (Chap. 9), the implications of these findings for fundamental research and for haptic applications will be discussed.

References

1. Abbink DA, Mulder M, Boer ER (2012) Haptic shared control: smoothly shifting control authority? Cogn Tech Work 14(1):19–28
2. Bergmann Tiest WM, Kappers AML (2014) Physical aspects of softness perception. Springer, London, pp 3–15
3. De Jonge A, Wildenbeest J, Boessenkool H, Abbink D (2016) The effect of trial-by-trial adaptation on conflicts in haptic shared control for free-air teleoperation tasks. IEEE Trans Haptic 9(1):111–120
4. Durlach N, Delhorne L, Wong A, Ko W, Rabinowitz W, Hollerbach J (1989) Manual discrimination and identification of length by the finger-span method. Percept Psychophys 46(1):29–38
5. Hale KS, Stanney KM (2004) Deriving haptic design guidelines from human physiological, psychophysical, and neurological foundations. IEEE Comput Graph Appl 24:33–39
6. Hayward V, Astley OR, Cruz-Hernandez M, Grant D, Robles-De-La-Torre G (2004) Haptic interfaces and devices. Sens Rev 24(1):16–29
7. Heit E (2000) Properties of inductive reasoning. Psychon Bull Rev 7(4):569–592
8. Jones LA, Tan HZ (2013) Application of psychophysical techniques to haptic research. IEEE Tran Haptics 6(3):268–284
9. Kimmig A (2013) Deductive reasoning. Springer, New York, pp 557–558
10. Klatzky RL, Lederman SJ (2003) Touch, vol 4. John Wiley and Sons, Inc., New Jersey, chap 6, pp 147–176
11. Lederman SJ, Jones LA (2011) Tactile and haptic illusions. IEEE Trans Haptics 4:273–294
12. Lederman SJ, Klatzky RL (2009) Haptic perception: a tutorial. Atten Percept Psychophysics 71(7):1439–1459
13. Loomis JM, Lederman SJ (1986) Tactual perception, chap 31. In: Boff K, Kaufman L, Thomas J (eds) Handbook of perception and human performance volume II: cognitive processes and performance. John Wiley and Sons, New York
14. Marayong P, Okamura AM (2004) Speed-accuracy characteristics of human-machine cooperative manipulation using virtual fixtures with variable admittance. Hum Factors J Hum Factors Ergon Soc 46(3):518–532
15. Nitsch V, Färber B (2013) A meta-analysis of the effects of haptic interfaces on task performance with teleoperation systems. IEEE Trans Haptics 6(4):387–398
16. Pierce R, Kuchenbecker K (2012) A data-driven method for determining natural human-robot motion mappings in teleoperation. In: 4th IEEE RAS EMBS international conference on biomedical robotics and biomechatronics (BioRob), pp 169–176
17. Proske U, Gandevia SC (2012) The proprioceptive senses: their roles in signaling body shape, body position and movement, and muscle force. Physiol Rev 92(4):1651–1697
18. Sigrist R, Rauter G, Riener R, Wolf P (2013) Augmented visual, auditory, haptic, and multimodal feedback in motor learning: a review. Psychon Bull Rev 20(1):21–53
19. Soechting JF, Flanders M (1989) Sensorimotor representations for pointing to targets in three-dimensional space. J Neurophysiol 62(2):582–594
20. Srinivasan MA, Basdogan C (1997) Haptics in virtual environments: taxonomy, research status, and challenges. Comput Graph 21(4):393–404

21. Stanney K (1995) Realizing the full potential of virtual reality: human factors issues that could stand in the way. In: Proceedings of the Virtual Reality Annual International Symposium (VRAIS '95), pp 28–34

22. Stevens S, Stone G (1959) Finger span: ratio scale, category scale, and JND scale. J Exp Psychol 57(2):91–95

23. Symmons M, Richardson B, Wuillemin D (2004) Active versus passive touch: superiority depends more on the task than the mode. In: Ballesteros S, Heller M (eds) Touch, blindness, and neuroscience. UNED Press, Madrid, pp 179–185

24. Tan HZ, Pang XD, Durlach NI (1992) Manual resolution of length, force and compliance. In: Kazerooni H (ed) Advances in robotics, vol 42. The American Society of Mechanical Engineers, pp 13–18

25. Van der Helm PA (2010) Weber-Fechner behavior in symmetry perception? Atten Percept Psychophysics 72(7):1854–1864

26. Vicentini M, Botturi D (2010) Perceptual issues improve haptic systems performance. In: Zadeh MH (ed) Advances in haptics. InTech, Vukovar

27. Weber E (1978/1834) De tactu. In: E.H. Weber on the tactile senses. Erlbaum (UK) Taylor & Francis, Hove

Part I
Static Perception

Chapter 2
Perception of Force Direction and Magnitude

Abstract Although force-feedback devices are already being used, the human ability to perceive forces has not been documented thoroughly. The haptic perception of force direction and magnitude has mostly been studied in discrimination tasks in the direction of gravity. In our study, the influence of physical force direction on haptic perception of force magnitude and direction was studied in the horizontal plane. Subjects estimated the direction and magnitude of a force exerted on their stationary hand. A significant anisotropy in perception of force magnitude and direction was found. Force direction data showed significant subject-dependent distortions at various physical directions. Normalized force magnitude data showed a consistent elliptical pattern, with its minor axis pointing roughly from the subject's hand to his/her shoulder. This pattern could be related to arm stiffness or manipulability patterns, which are also ellipse-shaped. These ellipses have an orientation consistent with the distortion measured in our study. So, forces in the direction of highest stiffness and lowest manipulability are perceived as being smaller. It therefore seems that humans possess a 'sense of effort' rather than a 'sense of force', which may be more useful in everyday life. These results could be useful in the design of haptic devices.

Previously published as:
F.E. van Beek, W.M. Bergmann Tiest & A.M.L. Kappers (2013)
Anisotropy in the haptic perception of force direction and magnitude
IEEE Transactions on Haptics, 6(4), 399–407
DOI:10.1109/TOH.2013.37.

2.1 Introduction

The application of haptic devices in teleoperation systems is increasing, but the presented force feedback does not always feel intuitive [19, 22]. So, to improve this, it would be useful to know more about the human perception of force. The perception of weight was the first subject investigated in haptic research, described in Weber's classic work on haptic weight perception [37]. When the development of haptic devices began, research changed from looking only at force perception in the direction of gravity [5, 13, 28] to looking at it in three dimensions [e.g. 10, 26, 36, 39]. This also changed the term 'weight perception' to 'force perception'. Both aspects of force, direction and magnitude, have mainly been investigated in discrimination experiments, focusing on the precision of force perception. The aim of our study was to investigate the relation between physical and perceived forces at different force directions and magnitudes, in order to obtain more insight into what humans are actually perceiving rather than with which precision they are perceiving it. The next section provides a summary of the literature on perception of force magnitude and direction.

The discrimination of **force magnitude** in the direction of gravity is a thoroughly investigated subject (for a review of force and weight perception, see Jones [17]). An important parameter in discrimination experiments is the Weber fraction, which refers to the minimum percentage of stimulus difference that is perceivable and describes the *precision of force magnitude perception*. Typically, Weber fractions range from 8% [6] to 13% [27] for force magnitude discrimination tasks in the direction of gravity. In contrast to this large body of studies, magnitude discrimination experiments in other directions are still scarce. A start was made by establishing discrimination thresholds for forces exerted on a stylus, using several direction-magnitude combinations in the fronto-parallel plane [26]. A significant anisotropy was found, as Weber fractions were about 10% in the upward, and 22% in the lateral direction. The diagonally right-up direction showed the highest Weber fraction of 30%. Unfortunately, the stimuli were forces that varied in direction and magnitude of force simultaneously, which makes it impossible to distinguish between the two. A similar method was used in another study [39], which tested force magnitude discrimination by providing forces while subjects were moving the stylus of a haptic device. They found a higher Weber fraction for forces at a 45° angle with the movement direction, compared to directions parallel and perpendicular to the movement. Together, these studies seem to suggest discrimination is poorer at some non-cardinal directions, with respect to the orientation of the hand, than at cardinal directions.

In addition to these studies, some more elaborate work on discrimination of force magnitude at cardinal directions has been done. Dorjgotov et al. [10] found a mean Weber fraction of 13% for forces exerted along all three cardinal axes. No discrimination difference between the axes was found. A similar result was found using a range of force magnitudes in a discrimination task where subjects were holding a handle of a haptic device (Weber fraction: 15%) [36]. So, for the cardinal directions the mean Weber fraction for force magnitude seems to be 13–15%.

Another parameter that can be investigated using discrimination experiments is the difference between force magnitude perception at different directions. This parameter, anisotropy, has been studied in only one experiment [10]. Subjects were asked to discriminate between a reference force, exerted along the dorso-ventral axis towards themselves, and a test force exerted on one of the cardinal axes, using a haptic device held with the whole hand. They perceived the magnitude of forces along the dorso-ventral axis towards themselves as larger than forces exerted in any other cardinal direction. There was no difference between the other directions tested.

A more direct way to investigate anisotropy is by studying the *relation between physical and perceived magnitude* through force magnitude estimation experiments. This would allow a comparison of force magnitude perception in various directions at various force levels. Only one study has investigated this, showing an anisotropy in force magnitude perception at different force directions [31]. However, they used a force range between 5 and 30 N, which are quite large forces when controlling a master device for a long time. The importance of stimulus direction for its perceived intensity has been hinted on by Dorjgotov et al. [10]. This effect has also been established in other haptic studies. For instance, the radial-tangential illusion is not only a well-known visual illusion, but also a haptic illusion of distance [1]. Subjects overestimate a distance that is presented radially, compared to a distance that is presented tangentially. As this is a very consistent effect, force magnitude perception could also be subject to such distortions.

The discrimination of **force direction** has also been studied in some experiments [3, 30]. In these, force direction discrimination tasks were performed to establish the *precision of perception of force direction* by exerting a force in the fronto-parallel plane on the passive index finger. This research was extended by using the same paradigm to test force direction perception of the index finger during active movement of the arm [38]. In all three studies, a JND (Just Noticeable Difference: the smallest absolute difference that is perceivable) of about 30° was found that did not differ significantly between reference directions. However, Elhajj et al. [11] found an influence of physical direction on JND for forces exerted in the horizontal plane in a discrimination experiment. Stimuli at every degree in the horizontal plane away from the subject were used. Afterwards, three regions of 60° each were defined. For each region, the percent-correct value was calculated. This showed that the medial region had a higher percentage of correctly discriminated trials than the left and right lateral regions.

One way to investigate the *relation between physical and perceived force direction* is by performing a matching task. Toffin et al. [33] are the only ones who have performed such an experiment. In their study, subjects had to reproduce a perceived force direction in the horizontal plane by moving a joystick, thus performing a perception task directly followed by a motor task. Different reproduction-errors for different directions were found, indicating an anisotropy in the reproduction of force direction. They did not report specific values for the different directions, but only tested the anisotropy as a whole. Moreover, their setup did not only test perception, but also motor performance as a response to perception. When there is a mismatch between the direction in which a force is exerted and the perceived direction, this mismatch could also be present in the perception of the reproduced force direction. This would result in two opposite errors that cancel out each other, so no error will be found. Therefore, it would be very useful to perform an experiment that involves only perception without a consecutive motor task to establish whether also in that case there is an anisotropy in the perception of force direction.

From the studies described, it is clear that literature on force perception in any direction other than the direction of gravity is limited and work on the relationship between physical forces and perceived forces in different directions is even more scarce. The aim of our study was therefore to investigate the perception of both magnitude and direction of forces at various non-cardinal directions in the horizontal plane, using a paradigm involving force magnitude estimation and force direction matching. Force magnitudes that are small enough to be used by operators in daily practice were used. Our study provides more insight into which direction and magnitude of forces humans are actually perceiving, rather than which differences they are able to perceive. It also shows if anisotropies, suggested by Dorjgotov et al. [10] and Toffin et al. [33] and in line with the illusion in Armstrong and Marks [1], are present in the haptic perception of force. This could eventually aid in the design of haptic devices with more intuitive force feedback.

2.2   Materials and Methods

2.2.1   Subjects

Ten right-handed (assessed using the Coren-test for handedness [7], naive subjects participated in this study, 4 male and 6 female, aged 22 ± 3 years (mean $\pm$ standard deviation), height 1.78 ± 0.08 m, with no known neurological disorders. All subjects gave written informed consent, received a compensation of 10 euros per hour and prior to the experiment were given written instructions on how to perform it. One additional subject had to be excluded from the analysis, because she was not able to perform the task correctly.

2.2.2   Setup

Subjects were seated on a height-adjustable chair (Fig. 2.1). Vision was blocked by a computer screen in front of the subject's eyes, enclosed in a tent-like structure that prevented visual cues from the side. The subject's elbow was put in a sling, attached to the test frame. The height of the chair and the distance between the chair and the setup was adjusted so that the arm posture was the same for every subject. By measuring the angles between the limbs, a vertical angle of 75° between torso and upper arm, and an angle of 130° between upper and lower arm was ensured for all

Fig. 2.1 Overview of the setup. To exert force on the handle, water was poured in the container through the funnel. The whole setup was turnable around the vertical axis to create different force directions in the horizontal plane. Joint angles were fixed by restraining the elbow in a sling, which was connected to the test frame. Hand position was controlled by letting subjects place the handle at the rest position between trials, to ensure every trial started at the same position

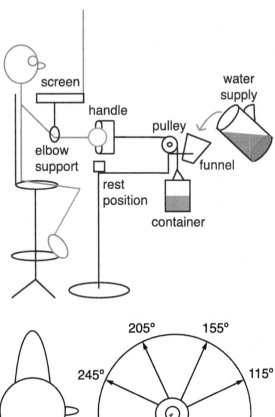

Fig. 2.2 Top view of the setup with used force directions for the right hand. For the left hand, mirrored directions were used. a: Angle between the upper and lower arm, set to 130° for all subjects; b: Angle between the line connecting shoulder and hand and the dorso-ventral axis of subject. This angle was roughly 25°

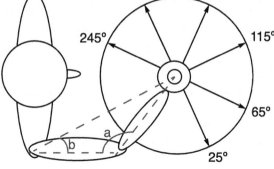

subjects (see Fig. 2.2 for a top view of the subject's position). These joint angles provided a comfortable posture, mimicking the posture of operators using a master device in real situations. In the resting phase between all trials, subjects placed the handle at the position indicated with 'rest position' in Fig. 2.1. The next trial started by letting the subjects lift the handle vertically, to ensure that all subjects performed the task using similar arm postures.

Subjects were asked to hold a handle on which a force could be exerted in the horizontal plane. The handle was connected to a container with a rope guided over a pulley. The stimulus force was increased gradually by pouring water in the container via a funnel, creating a force-ramp that ended at a certain plateau-force when all the water had reached the container. The steepness of the force ramp was controlled by changing the size of the opening of the funnel. To exert force in different directions, the complete pulley-funnel system was turnable. The handle also had a turnable inside that could rotate independent from the handle's outside, to allow changing the direction of the setup without the subject noticing in which direction the setup was turning.

2.2.3 Experimental Procedure

During every trial, the force on the handle was gradually increased to a certain force-plateau through a force-ramp. By adjusting the funnel diameter, the force-ramp lasted 5 s for all force magnitudes. Subjects were wearing ear protectors with ear phones throughout the experiment. During the force-ramp, white noise was played to mask the sound of the water. Subjects were free to choose how long they wanted to perceive the plateau-force before answering, generally resulting in a few seconds of plateau-force exposure. The subject's task was to answer two questions about his/her perception of the force at this plateau-level, indicating the direction and the magnitude of the perceived force. Perceived direction was indicated by turning an arrow on a computer screen, placed in the horizontal plane beneath the subject's face. The arrow was manipulated using a computer mouse in his/her free hand by either pointing and clicking on the screen or by dragging the needle to the desired position. When the subjects were satisfied with their answer, they pressed an 'ok'-button to confirm their response.[1] Magnitude perception was indicated verbally through free magnitude estimation. No magnitude reference or range was provided.

Six directions covering three-quarters of the horizontal plane (245, 205, 155, 115, 65 and 25°, see Fig. 2.2 for an overview of these directions) and five magnitudes (2, 3, 4, 5 and 6 N) were tested, resulting in 30 combinations for each hand. All direction-magnitude combinations were tested six times: three times while subjects held the handle with their right hand and three times with their left hand. The experiment was divided into three 1-h test sessions per subject. At each session, all 30 direction-magnitude combinations were tested once on both, resulting in a total of 180 trials per subject. The hand-order was counterbalanced and the direction-magnitude combinations were presented in a randomized order that was different for every session. Every session started with three practice trials to familiarize the subjects with the task. If the subjects did not perform the task correctly, they were given feedback to adjust their procedure. Generally, this was only needed during the practice trials. Subjects did not receive feedback about the responses they gave.

---

[1] We are aware that this matching procedure could be influenced by visual distortion. However, there is no method that guarantees a bias-free measurement of these parameters.

2.2.4   Data Analysis

Coordinates of the left-hand trials were mirrored before analysis to ensure that the same coordinates represented the same directions with respect to the subject's hand in both data sets, i.e. all data were represented in right-hand coordinates. Perceived force directions were directly compared to physical force directions. Literature on magnitude perception in the direction of gravity reported power functions with different exponents for the relation between physical and perceived magnitude [17]. Since this exponent is needed to choose the proper normalization method, a power function was fitted to our magnitude perception data at the different levels of physical force magnitude. This was done for every physical direction separately, for the individual values (used in the statistical analysis) and for the data of all subjects together (Fig. 2.3 and Table 2.1). At one of the angles the exponents were not normally distributed over the subjects (Shapiro-Wilkinson test, $D_{10} = 0.82$, $p = 0.028$), so a non-parametric test was used to assess the medians of the exponents based on the individual fits. As the fits did not differ from a linear fit, as will be described in more detail in the results section, a simple normalization method could be used. This normalization was done by dividing the perception data by the actual force magnitude, resulting in values that represented the ratio between perception and actual force magnitudes.

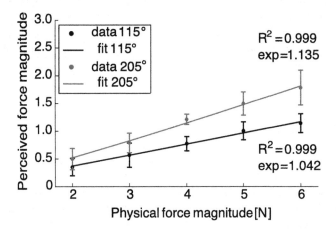

Fig. 2.3 Scaling of mean force magnitude perception data of all subjects as a function of physical force magnitude. The black dots show the magnitude data of the 115° direction, the grey dots show the data of the 205° direction, and error bars indicate standard deviations. A power function was fitted to the data (the solid black and grey lines). The fit showed a very high correlation with the data and its exponent did not differ from 1, which indicates a linear relationship. For clarity, only two of the six curves are shown. Note the difference in slope between the two lines, indicating a difference in force magnitude perception between the two directions

Table 2.1 Exponents of all power functions, fitted to the magnitude perception data per subject per physical direction. The last two rows show the exponents and the coefficients of determination, R^2, of the power functions fitted to the data of all subjects

Subject	25°	65°	115°	155°	205°	245°
1	1.09	0.97	1.37	1.44	1.05	1.05
2	0.96	1.06	0.93	0.96	1.02	0.86
3	0.89	1.05	0.92	1.52	1.08	1.16
4	0.78	0.64	0.75	0.75	0.78	0.80
5	0.88	0.84	1.50	1.02	0.99	1.27
6	0.87	1.07	1.00	1.04	1.21	1.17
7	1.35	1.00	0.87	1.18	1.35	1.27
8	1.10	1.24	1.29	1.62	1.30	1.30
9	1.73	1.97	2.40	3.23	2.05	2.29
10	0.39	0.46	0.25	0.42	0.41	0.38
All	1.02	1.03	1.04	1.17	1.14	1.13
R^2 all	0.9991	0.9986	0.9993	0.9997	0.9990	0.9997

From these ratio values, the mean was calculated per subject per session. The ratios were then scaled to have a mean of one, by dividing the ratio values by the mean, per subject and session. This resulted in ratios that all had a mean of one per subject and session, but which still represented the distribution of values before normalization and scaling. After this, an outlier analysis was performed on both the normalized magnitude and the direction data per subject. This resulted in a maximum number of nine outliers per subject (maximally 5% of the subject's data set, but for most subjects about 2%). Means of all magnitude-direction combinations were calculated by averaging the data from the three sessions.

2.2.5 Statistics

The effects of physical force direction, physical force magnitude and used hand on the errors in direction and normalized magnitude values were analyzed with a repeated measures ANOVA. When the sphericity-criterion was not met, Greenhouse-Geisser correction was used to adjust the degrees of freedom. To assess if the medians of the errors in direction were different from zero, a non-parameteric Kruskal-Wallis test per subject per physical angle was performed, as the individual data per angle were not normally distributed (Kolmogorov-Smirnov test, $D_{252} \leq 0.13$, $p \leq 0.022$, for all physical directions). An ellipse was fitted to the normalized magnitude data per subject and for all the data together. The mean eccentricity of the ellipses was tested using a one-sided one-sample t-test, as these parameters were normally distributed (Shapiro-Wilkinson test, $D_{10} = 0.86$, $p = 0.15$). Eccentricity values cannot exceed the value 1, therefore a one-sided test was used. The consistency of the error patterns for force direction were assessed using Pearson's correlation coefficient, r, along with its significance. Correlations were calculated per subject by comparing the error patterns of the different force magnitudes, hands and sessions.

2.3 Results

2.3.1 Force Magnitude

Scaling of force magnitude perception was tested by fitting a power function to the data. A Wilcoxon Signed Rank test showed for all directions that the median of the exponents did not differ significantly from 1 ($0.24 \leq p \leq 0.88$), so the scaling of force magnitude data did not differ from linear. In Fig. 2.3 and Table 2.1, the relation between physical and perceived force and the fit is plotted, showing that the ratio of physical to perceived force was constant over the force range. Repeated measures ANOVA on normalized magnitude values showed a significant effect of physical angle ($F_{2.3,20} = 19$, $p < 0.01$), but no effect of physical force magnitude ($F_{1.1,9.6} = 1.1$, $p = 0.32$) or used hand ($F_{1,9} = 0.39$, $p = 0.55$). Consequently, the data set for magnitude could be merged for hand and physical force magnitude.

When the normalized magnitude data were plotted in a polar plot, the data resembled an elliptical pattern (Fig. 2.4a). Ellipses were fitted to the data for every subject and to the complete data set of all subjects together (Fig. 2.4b), which fitted the data quite well. This can be seen from the coefficient of determination, R^2, which ranged between 0.82 and 0.95 for the individual fits. Data of one subject had a somewhat poorer fit of 0.61. The coefficient of determination for the fit to the complete data set was 0.86. The eccentricity of the ellipses was significantly smaller than 1 (one-sided one-sample t-test: $t_9 = -7.6$, $p < 0.01$), meaning that the fitted figure is an ellipse and not a circle (Fig. 2.5a). The orientation of the ellipses was very similar over subjects (Fig. 2.5b). The eccentricity of the ellipse fitted to the complete data set was 0.66 and the orientation of its major axis was 52°.

2.3.2 Force Direction

From the repeated measures ANOVA of the differences between physical and perceived direction (see Fig. 2.6a for an overview of all force direction data), it appeared that there was only an effect of physical force direction ($F_{2.4,22} = 4.5$, $p = 0.018$) and none of physical force magnitude ($F_{2.0,18} = 1.4$, $p = 0.27$) or used hand ($F_{1,9} = 0.24$, $p = 0.64$). Therefore, the direction data set was merged for hand and physical force magnitude. A Kruskal-Wallis test showed that all subjects made significant errors at various physical directions. In Fig. 2.6a, the filled dots

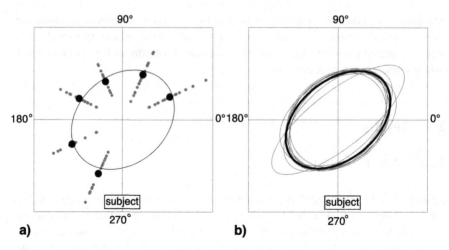

a)   **b)**

Fig. 2.4 Magnitude perception data with fitted ellipses, showing a polar view of the magnitude perception data at different physical directions. The greater the distance from the centre of the graph, the higher the perceived magnitude was. Data from the left-hand trials were mirrored before averaging over hands. The subject was sitting at the marked position. (**a**) Data of one typical subject. Grey small dots represent individual trials, while the large black dots show the mean magnitude perception values in that direction. An ellipse was fitted through these data, shown in black. Note that although the spread of the data is quite large, the general elliptical shape of the data is very apparent. (**b**) Ellipses fitted to normalized magnitude perception data for the individual subjects (grey) and the mean fit for the complete data set (black)

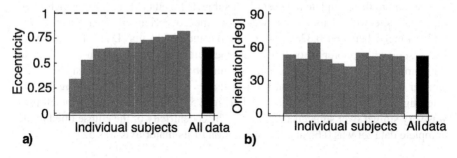

a)   **b)**

Fig. 2.5 Parameters of ellipse fit, for individual subjects (gray) and for the fit to the complete data set (black). (**a**) Eccentricity of the fitted ellipses. The dotted line indicates the value 1, which would be a perfect circle. (**b**) Orientation of the major axes of the fitted ellipses. Note that especially the ellipse orientation varies very little between subjects

indicate these significant differences, showing that errors differed from zero for 48 of the 60 subject-direction combinations. There was, however, a large range in errors between subjects. Moreover, the error patterns did not have the same shape, which is clear from the lines connecting the errors per subject in Fig. 2.6a. An alternative representation of an error pattern is given in Fig. 2.6b, by showing the errors of a subject in a top view of the set-up. Although the patterns were not consistent

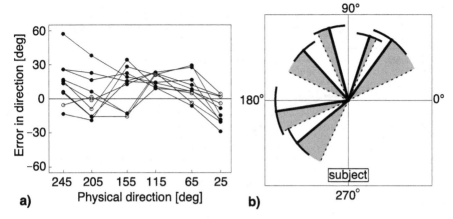

Fig. 2.6 Difference between the physical and perceived force directions. Data from the left-hand trials were mirrored before averaging over hands. (**a**) Errors made by the subjects, ranging from the palmar side of the hand (left part of the graph) to the dorsal side of the hand (right part of the graph). Each dot represents the error of one subject at that physical direction and each line connects all errors of one subject. Filled (open) dots show physical directions at which the subject made an error that was (not) significantly different from 0. The range of errors was large and the shape of error patterns differed between subjects. (**b**) Alternative representation of an error pattern of one subject, shown in a top view of the set-up, with the subject sitting at the marked position. Dotted lines represent physical angles, solid lines represent perceived angles, shaded areas show the difference between these two (i.e. the error subjects made) and arcs show the standard deviation. Reading the graph in (**a**) from left to right is congruent to reading the graph in (**b**) in a clockwise direction

between subjects, the consistency of the error patterns within subjects was quite strong, as illustrated in Fig. 2.7. This figure illustrates the general trend that within the data of one subject, the same patterns arise when the errors in direction are compared between the different force magnitudes, hands and sessions. Correlation analysis of these comparisons per subject confirmed that the patterns were quite consistent, as 53 of the 100 magnitude correlations, 5 of the 10 hand correlations, and 23 of the 30 session correlations were significant. The median value of the coefficients was 0.84 for the force magnitudes, 0.85 for the hands and 0.92 for the sessions.

2.4 Discussion

Since for both the perception of force magnitude and the perception of force direction a significant effect of physical force direction was found, force perception in the horizontal plane is anisotropic. This is an important finding with regard to applications of haptic force perception, such as haptic force-feedback control systems. Of course, it is also an interesting finding in a more fundamental sense.

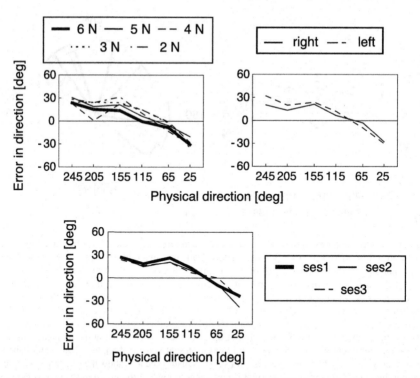

Fig. 2.7 Illustration of the within-subject consistency of errors in direction perception, using data of one typical subject. Top left: error pattern per physical force magnitude, with one line per magnitude. Top right: error pattern per used hand, with one line per hand. Bottom: error pattern per session, with one line per session. All sessions were measured on different days, which makes this an indicator of the repeatability of the measurements over time. The error patterns are consistent over physical force magnitudes, used hand and measurement days

2.4.1 Force Magnitude

Scaling of force magnitude did not differ from linear for the magnitude range in this study, so the ratio of perceived to physical magnitude was constant. This agrees with results of a study in which the scaling of force, ranging between 0.15 and 0.70 N and applied normally and tangentially to the index finger, was linear for both force directions [23]. Earlier studies on force magnitude scaling have mainly used forces in the direction of gravity or magnitudes that were much higher. In these studies, power functions with exponents varying between 0.7 and 2.0 were found [17, 31], so our data fit in that range. No influence of physical force magnitude was found, so the different force magnitudes did not cause significantly different normalized force magnitude perception patterns.

The pattern of normalized force magnitude data in a polar plot showed a remarkably similar-oriented elliptical pattern for all subjects (see Fig. 2.4b). The minor axis of the ellipse was always oriented roughly in the same direction as the arm (Fig. 2.5b). This indicates that subjects perceived a force as larger when it was exerted perpendicular to the arm, than when it was exerted along the arm. The observed pattern makes sense intuitively, as you would expect that resisting a force in line with the arm is easier than resisting one perpendicular to the arm. Nonetheless, it is intriguing that it is apparently not only easier to do, but is also perceived as a lower force magnitude. The way in which an arm reacts to external disturbances can be described using arm impedance characteristics [35]. For small displacements and without voluntary muscle control, the following equation describes arm impedance:

$$M\ddot{\mathbf{x}}(t) + B\dot{\mathbf{x}}(t) + K\mathbf{x}(t) = \mathbf{f}(t). \tag{2.1}$$

in which M, B and K indicate the matrices of inertia, viscosity and stiffness, respectively, and $\mathbf{f}$ indicates the force driving the arm to move.

All of these parameters are anisotropic for the human arm (e.g. [32]) and can be represented with an ellipse, which suggests a connection with our perception data. The shape and orientation of these ellipses depend on many factors, among which arm position is very important [8, 9]. In our study, we tried to ensure that all subjects used the same posture. The chair could not turn, elbow position was fixed, the start position of the hand was the same in every trial and the subjects were instructed to keep their hand at the same place during the trials. However, we did not measure arm position directly, so we cannot calculate the absolute contribution of the different parameters. Nonetheless, we can make an educated guess about their relative contribution, based on qualitative analysis of the parameters. Artemiadis et al. [2] asked subjects to keep their hand steady while a dynamic force profile was applied and found that the stiffness ellipse (K) was magnitudes larger than the viscosity (B) and inertia (M) ellipse. In our study, the task was also to keep the hand steady. Some horizontal motion ($\mathbf{x}$) was observed during the trials, but the displacement was a few centimeters maximally, as the setup did not allow for larger movements. If movement was observed, this happened very slowly, because the force ($\mathbf{f}$) was added gradually. Consequently, the velocity ($\dot{\mathbf{x}}$) and acceleration ($\ddot{\mathbf{x}}$) values were probably relatively small. Taken together, it seems unlikely that the products $M\ddot{\mathbf{x}}$ and $B\dot{\mathbf{x}}$ played a large role in the total arm impedance, making stiffness the most likely governing arm impedance parameter in our study.

For tasks avoiding voluntary muscle control and performed with the arm in the horizontal plane, the arm stiffness ellipse is oriented along the line between hand and shoulder [12, 15, 16, 21, 35]. When voluntary muscle control is present, as was the case in our study, it can change the size and the eccentricity of the stiffness ellipse [8] and its orientation [8, 9, 24, 25]. However, Krutky, Trumbower an Perreault [18] show that the orientation of the ellipse does not change in a task where subjects have to maintain a posture, while an external disturbance is presented, which was also done in our study. Consequently, the orientation of the stiffness ellipse measured without voluntary muscle force will be good enough as a rough estimation of the orientation of the stiffness ellipse in our study, even though voluntary muscle control was present. To provide a natural posture for task execution, the arm of our subjects was not positioned in the horizontal plane. One other study investigating arm stiffness outside of the horizontal plane used a posture with the elbow below the shoulder and hand [34]. The shoulder and hand were both positioned in the horizontal plane. In this posture, the major axis of the arm stiffness ellipse was still oriented from hand to shoulder in the horizontal projection. Therefore, we assume that maximum arm stiffness for the posture in our study was also oriented along the hand-shoulder line.

A link between arm impedance and motor behaviour was found by showing that arm stiffness ellipses were correlated to the very consistent elliptical anisotropy in forces exerted by subjects in a motor task [29]. In our study, magnitude perception of the different subjects also showed a very consistent elliptical anisotropy. The exact orientation of the hand-shoulder line, and thus of the hypothesized major axis of the stiffness ellipse, was not documented in our study, but we estimated the orientation to be 25° on average, based on measurements of arm length, elbow angle and setup configuration (see Fig. 2.2 for this hand-shoulder line). In Fig. 2.8 both the fitted mean magnitude perception ellipse and the estimated hand-shoulder line are shown. The major axis of the arm stiffness ellipse from Mussa-Ivaldi et al. [21] with a posture most similar to the posture in our study is also shown. From Fig. 2.8 it is clear that the orientation of the hand-shoulder line in our study and the stiffness data from literature [21] show a remarkable similarity to our measured force magnitude data. So, both our educated guess and the general features of arm stiffness suggest this parameter could be the governing one.

Our force perception results fit nicely with a study by Tanaka and Tsuji [31], who looked at force magnitude perception at larger magnitudes. They also found elliptical force perception patterns, oriented roughly in the same direction as our ellipses. Their explanation for the perception anisotropy is based on another arm characteristic, which is arm manipulability. Arm manipulability refers to the ease with which an external force can displace an arm, either robotic or human, in a certain direction, based on arm configuration and the range of possible joint velocities or torques [2, 4, 32]. Tanaka and Tsuji [31] argue that the ellipses they find for force perception are very similar to the manipulability ellipses of the human arm. The minor axes of the manipulability ellipses in their study are oriented along the line between hand and shoulder. So, both the stiffness ellipse and the manipulability ellipse point roughly from hand to shoulder, as can been seen in Fig. 2.8. They both show a remarkable similarity to the force perception patterns we found and both give a measure of the ease of force production in a certain direction. So, regardless of which parameter is the main factor, the results suggest that static force magnitude perception in the horizontal plane is governed by the ease with which the force is resisted rather than the actual force magnitude.

This agrees with theories in literature, stating that humans possess a 'sense of effort', describing the sense of ease of force production. This term was introduced by McCloskey et al., when they found that subjects who fatigued their reference

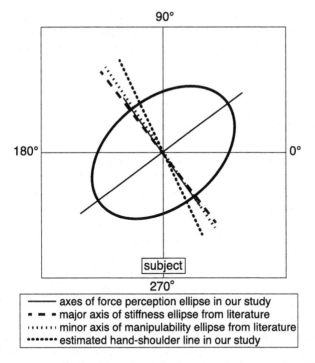

Fig. 2.8 Possible explanation for ellipses found for force magnitude perception. The fitted mean force magnitude ellipse to our complete data set is shown, with its major and minor axis as a solid line and the position of the subject marked. The thick dashed line indicates the major axis of the arm stiffness ellipse found in Mussa-Ivaldi et al. [21] (hand-shoulder angle 35°), the thin dashed line indicates the arm manipulability ellipse found in Tanaka and Tsuji [31] (hand-shoulder angle 35°) and the dotted line indicates the estimated orientation of the line between hand and shoulder in our study (roughly 25°). All the segmented lines are very similar to the orientation of the minor axis of the force magnitude perception ellipse found in our study, so both stiffness and manipulability could be related to force magnitude perception

arm by keeping a weight lifted, chose higher matching weights over time. The matching weights were held for a short period, which avoided fatigue in the matching arm [20]. The same effect is observed when forces are perceived with locally anesthetized hands or when forces are perceived by patients with particular neuromuscular disorders (e.g. [14]). All these studies show that even when the force does not change, the perception of force does change according to the amount of effort subjects experience. This 'sense of effort' is probably related to kinaesthetic perception rather than tactual perception, as it is related to arm mechanics. However, we cannot conclude this from our experiment, as a combination of kinaesthetic and tactile cues was present.

It is interesting that humans, who have been perceiving forces all their life, are not able to correct for the anisotropic nature of effort in order to obtain a correct sense of force. It is probably more instructive to have information about effort rather than absolute force, as effort represents the ease with which one will be able to perform a task. However, when force is used to communicate information, as in a haptic device, it might be better to scale the force magnitude according to its direction to produce an isotropic sense of effort.

2.4.2 Force Direction

Significant distortions in perception of force direction were found for all subjects. No influence of physical force magnitude was found, so the different force magnitudes did not cause significantly different force direction errors. Within subjects the patterns were quite consistent (Fig. 2.7 and the correlation analysis), but between subjects, large differences were found (Fig. 2.6a, b). As the mean errors in direction were significantly different from zero for multiple physical directions for all subjects, these errors are not just random variations, but indications of actual distortions. The presence of an effect of physical force direction on the error in perception shows that this distortion is not a simple rotation of the complete system, but differs per physical direction. If the errors were related to arm mechanics, all subjects would show a similar pattern, as their arm postures were similar. This is not the case, so stiffness or manipulability cannot be used to explain the variation found between subjects. More research is needed to answer this question. Nonetheless, the consistency of the patterns within subjects could already be useful in the design of force-feedback, by using the pattern particular to that operator to adjust the direction of the forces that are fed back.

Lastly, no effect of used hand on the perception of force direction or magnitude was found, even though all participants were right-handed, indicating that the ability to haptically perceive force is not trained by using hands more often or for more precise tasks. The dominant hand did not yield lower magnitude perception values. Probably, this happened because the trials with the same hand were presented in a block at every session. This allowed for a re-scaling of the perception data to the experienced force range when switching to the other hand half-way the session.

2.5 Conclusion

Both the perception of force direction and magnitude was anisotropic in the horizontal plane. Perception of force direction was significantly distorted at various physical directions for all subjects. Between subjects, the patterns of these errors varied. Within subjects, the patterns were quite consistent. Distortion of force magnitude showed an elliptical pattern that was very comparable between subjects. All subjects perceived forces exerted along the line between shoulder and arm to be smaller than forces exerted perpendicular to this line. The force magnitude ellipses were oriented roughly perpendicular to arm stiffness and similar to the arm manipulability ellipses found previously, meaning that forces perceived in a direction with higher arm stiffness and lower manipulability are perceived as being smaller. Humans thus seem to possess a 'sense of effort' rather than a 'sense of force', which might be more helpful in performing tasks of everyday life. Both the distortion in direction and magnitude of force perception are interesting phenomena that could be important for the design of haptic devices.

References

1. Armstrong L, Marks L (1999) Haptic perception of linear extent. Percept Psychophys 61(6):1211–1226
2. Artemiadis P, Katsiaris P, Liarokapis M, Kyriakopoulos K (2010) Human arm impedance: characterization and modeling in 3D space. In: International conference on intelligent robots and systems, pp 3103–3108
3. Barbagli F, Salisbury K, Ho C, Spence C, Tan HZ (2006) Haptic discrimination of force direction and the influence of visual information. ACM Trans Appl Percept 3(2):125–135
4. Bicchi A, Melchiorri C, Balluchi D (1995) On the mobility and manipulability of general multiple limb robots. IEEE Trans Robot Autom 11(2):215–228
5. Brodie EE, Ross HE (1984) Sensorimotor mechanisms in weight discrimination. Atten Percept Psychophys 36(5):477–481
6. Brodie EE, Ross HE (1985) Jiggling a lifted weight does aid discrimination. Am J Psychol 98(3):469–471
7. Coren S (1993) The left-hander syndrome. Vintage Books, New York
8. Darainy M, Malfait N, Gribble PL, Towhidkhah F, Ostry DJ (2004) Learning to control arm stiffness under static conditions. J Neurophysiol 92(6):3344–3350
9. Darainy M, Towhidkhah F, Ostry DJ (2007) Control of hand impedance under static conditions and during reaching movement. J Neurophysiol 97(4):2676–2685
10. Dorjgotov E, Bertoline G, Arns L, Pizlo Z, Dunlop S (2008) Force amplitude perception in six orthogonal directions. In: Proceedings of the symposium on haptics interfaces for virtual environment and teleoperator systems, pp 121–127
11. Elhajj I, Weerasinghe H, Dika A, Hansen R (2006) Human perception of haptic force direction. In: Proceedings of the IEEE/RSJ international conference on intelligent robots and systems, pp 989–993
12. Flash T, Mussa-Ivaldi F (1990) Human arm stiffness characteristics during the maintenance of posture. Exp Brain Res 82(2):315–326
13. Gandevia S, McCloskey D (1976) Perceived heaviness of lifted objects and effects of sensory inputs from related, non-lifting parts. Brain Res 109(2):399–401
14. Gandevia S, McCloskey D (1977) Sensations of heaviness. Brain 100(2):345–354
15. Gomi H, Kawato M (1997) Human arm stiffness and equilibrium-point trajectory during multi-joint movement. Biol Cybern 76(3):163–171
16. Hogan N (1985) The mechanics of multi-joint posture and movement control. Biol Cybern 52(5):315–331
17. Jones LA (1986) Perception of force and weight: theory and research. Psychol Bull 100(1):29–42
18. Krutky MA, Trumbower RD, Perreault EJ (2009) Effects of environmental instabilities on endpoint stiffness during the maintenance of human arm posture. In: Annual international conference of the IEEE engineering in medicine and biology society, pp 5938–5941
19. Kuchenbecker K, Fiene J, Niemeyer G (2006) Improving contact realism through event-based haptic feedback. IEEE Trans Vis Comput Graph 12(2):219–230

20. McCloskey D, Ebeling P, Goodwin G (1974) Estimation of weights and tensions and apparent involvement of a 'sense of effort'. Exp Neurol 42(1):220–232
21. Mussa-Ivaldi F, Hogan N, Bizzi E (1985) Neural, mechanical, and geometric factors subserving arm posture in humans. J Neurosci 5(10):2732–2743
22. O'Malley MK, Goldfarb M (2005) On the ability of humans to haptically identify and discriminate real and simulated objects. Presence Teleop Virt Environ 14(3):366–376
23. Pare M, Carnahan H, Smith A (2002) Magnitude estimation of tangential force applied to the fingerpad. Exp Brain Res 142(3):342–348
24. Perreault EJ, Kirsch RF, Crago PE (2001) Effects of voluntary force generation on the elastic components of endpoint stiffness. Exp Brain Res 141(3):312–323
25. Perreault EJ, Kirsch RF, Crago PE (2002) Voluntary control of static endpoint stiffness during force regulation tasks. J Neurophys 87(6):2808–2816
26. Pongrac H, Hinterseer P, Kammerl J, Steinbach E, Färber B (2006) Limitations of human 3D-force discrimination. In: Proceedings of the second international workshop on human-centered robotics systems
27. Raj D, Ingty K, Devanandan M (1985) Weight appreciation in the hand in normal subjects and in patients with leprous neuropathy. Brain 108(1):95–102
28. Ross HE, Gregory RL (1970) Weight illusions and weight discrimination—a revised hypothesis. Q J Exp Psychol 22(2):318–328
29. Shadmehr R, Mussa-Ivaldi F, Bizzi E (1993) Postural force fields of the human arm and their role in generating multijoint movements. J Neurosci 13(1):45–62
30. Tan HZ, Barbagli F, Salisbury K, Ho C, Spence C (2006) Force-direction discrimination is not influenced by reference force direction. Haptics-e 4:1–6
31. Tanaka Y, Tsuji T (2008) Directional properties of human hand force perception in the maintenance of arm posture. In: Ishikawa M, Doya K, Miyamoto H, Yamakawa T (eds) Neural information processing. Lecture notes in computer science, vol 4984. Springer, Berlin/Heidelberg, pp 933–942
32. Tanaka Y, Yamada N, Nishikawa K, Masamori I, Tsuji T (2005) Manipulability analysis of human arm movements during the operation of a variable-impedance controlled robot. In: IEEE/RSJ international conference on intelligent robots and systems, pp 1893–1898
33. Toffin D, McIntyre J, Droulez J, Kemeny A, Berthoz A (2003) Perception and reproduction of force direction in the horizontal plane. J Neurophys 90(5):3040–3053
34. Trumbower RD, Krutky MA, Yang BS, Perreault EJ (2009) Use of self-selected postures to regulate multi-joint stiffness during unconstrained tasks. PLoS ONE 4(5):e5411
35. Tsuji T, Morasso P, Goto K, Ito K (1995) Human hand impedance characteristics during maintained posture. Biol Cybern 72(6):475–485
36. Vicentini M, Galvan S, Botturi D, Fiorini P (2010) Evaluation of force and torque magnitude discrimination thresholds on the human hand-arm system. ACM Trans Appl Percept 8(1):1–16
37. Weber E (1978/1834) De tactu, Erlbaum (UK) Taylor & Francis, Hove. E.H. Weber On The Tactile Senses
38. Yang XD, Bischof W, Boulanger P (2008a) The effects of hand motion on haptic perception of force direction. In: Ferre M (ed) Haptics: perception, devices and scenarios. Lecture notes in computer science, vol 5024. Springer, Berlin/Heidelberg, pp 355–360
39. Yang XD, Bischof W, Boulanger P (2008b) Perception of haptic force magnitude during hand movements. In: Proceedings of the IEEE international conference on robotics and automation, pp 2061–2066

Chapter 3
Perception of Force Direction

Abstract In a previous study, we found that the accuracy of human haptic perception of force direction is not very high. We also found an effect of physical force direction on the error subjects made, resulting in 'error patterns'. In the current study, we assessed the between- and within-subject variation of these patterns. The within-subject variation was assessed by measuring the error patterns repeatedly over time for the same set of subjects. Many of these patterns were correlated, which indicates that they are fairly stable over time and thus subject-specific. The between-subject analysis, conversely, yielded hardly any significant correlations. We also measured general subject parameters that might explain this between-subject variation, but these parameters did not correlate with the error patterns. Concluding, we found that the error patterns of haptic perception of force direction are subject-specific and probably governed by an internal subject parameter that we did not yet discover.

Previously published as:
F.E. van Beek, W.M. Bergmann Tiest, F.L. Gabrielse, B.W.J. Lagerberg,
T.K. Verhoogt, B.G.A. Wolfs & A.M.L. Kappers (2014)
Subject-specific distortions in haptic perception of force direction
In M. Auvray & C. Duriez (Eds.), *Haptics: Neuroscience, Devices, Modeling, and Applications*
Part I (Vol. 8618 of *Lecture Notes in Computer Science*, pp. 48–54)
Berlin/Heidelberg: Springer-Verlag.

3.1 Introduction

Force feedback is a very important aspect of haptic devices. To understand how force feedback algorithms should be designed, it is useful to gain insight in the human perception of force. In this study, we focussed on the human perception of the direction of a force. Like with all measurements, the terms *precision* and *accuracy* can also be used in relation to perception. Precision (also called discrimination threshold or variability) refers to the random error that subjects make, so it indicates the spread of the data around the perceived mean value. Accuracy (also called bias) refers to the systematic error that subjects make, so it indicates if the perceived mean differs from the physical mean.

Some studies already evaluated the precision of the perception of the direction of a force exerted on the passive index finger in the fronto-parallel plane [1, 5, 6]. They report a precision of 30°, which was the same for all force directions. Switching from a stationary to a moving arm seems to have no influence for forces exerted in the fronto-parallel plane, as the precision stays the same [10]. This does not hold for the perception of force direction of a shear-force exerted on the finger tip, as in this case the precision is higher when the arm is stationary [9]. For forces in the horizontal plane, force direction does seem to influence the precision of perception [4], with the median region showing a higher precision than the left and right lateral areas.

While there is some knowledge of the precision of perception of force direction, work on the accuracy of perception of force direction is very scarce. Toffin et al. [7] investigated this topic by asking subjects to move the handle of a joystick in the direction of a force presented earlier. However, this only shows the accuracy of the motor output, which could be adjusted for biases in perception. To investigate the accuracy of perception itself, we asked people directly for their perception of force direction in a previous study [8]. We found that humans make large errors, and thus are quite inaccurate, in this task. We also found a significant effect of physical force direction on the error humans make in perceiving the direction of the force. This indicates that the errors are not constant nor random, but form a kind of 'error-pattern'. In the current study, an attempt was made to investigate what the basis of these patterns is and how consistent they are over time.

In experiment 1, the within-subject differences in patterns over time were investigated. This was done by measuring the error patterns of a small group of subjects at different moments in time. The aim of this experiment was to investigate if the patterns are stable over time and thus subject-specific. Therefore, a small number of subjects were studied, while many measurements were performed for each subject.

In experiment 2, the between-subject differences in error patterns were investigated. In addition to the error patterns, a number of general subject parameters, which might be correlated to the between-subject variation, was measured. The aim of this experiment was to find groups of subjects that show similar patterns and to investigate if general subject parameters might explain the differences between the (hypothetical) groups.

Together, these experiments show the intra- and inter-subject differences in the perception of force direction. This information might be useful in the design of haptic feedback algorithms, because force direction is used to communicate information in this application. To make sure that the user understands the information properly, it might be beneficial to adjust force direction according to the error pattern of the user.

3.2   Material and Methods

3.2.1   Subjects

In experiment 1, 8 male subjects participated, aged 22 ± 2 years (mean $\pm$ s.d.). In experiment 2, 21 male and 4 female subjects participated, aged 21 ± 2 years.

In both experiments participants were naive to the purpose of the experiment. They were all right-handed, which was assessed using a Coren-test for handedness [3]. All participants signed an informed consent form. They received no compensation for their time.

3.2.2   Set-Up

Subjects were seated in front of the set-up (see Fig. 3.1a) on a height-adjustable chair. Their vision was blocked by a black screen. Beneath their faces, in the horizontal plane, the screen of a laptop was placed. Subjects used their right hand to hold a handle and their left hand to control a computer-mouse to provide the answers. Even though they controlled the mouse with their non-dominant hand, they reported that this was not a difficult task. There was also no time limit on answering, so they could take the time they needed to perform the task precisely. At each trial, subjects were prompted via the screen to lift the handle from the

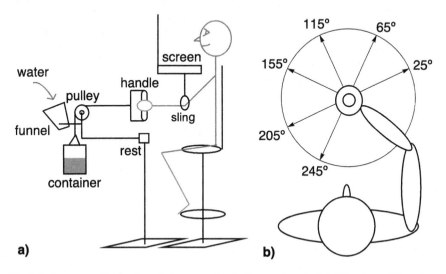

Fig. 3.1 Diagrams of the set-up that was used in both experiments. (**a**) Side view of the set-up showing the apparatus designed to deliver forces at different directions in the horizontal plane. (**b**) Top view of the same set-up, to indicate the force directions that were used in both experiments

resting position and then keep their hand at the same position in the air throughout the trial. At this point, the experimenter gradually increased the force by pouring water in the container, which was connected to the handle. The gradual increase (4 N was reached after 5 s) minimized inertial cues and made sure that subjects could keep their hand at the same position. Moreover, the funnel made sure that the force ramp was similar in all trials. During this force-ramp, white noise was played on the headphones, which the subjects were wearing. Once all the water was in the container and thus the force had reached its plateau-level of 4 N, the noise was switched off and the subjects answered the question: which direction do you think the force is coming from? They did this by turning the needle of a gauge shown on the screen, followed by a button-press to confirm their answer. The water was then removed from the container and the handle was returned to the resting position. To provide different force directions—which were 25°, 65°, 115°, 155°, 205° and 245° in both experiments, as shown in Fig. 3.1b—the set-up was turnable around its vertical base.

3.2.3 Protocol

In experiment 1, subjects performed at least three experimental sessions of 1 h each within 1 week, on three different days. During each session, every force direction was presented ten times in a random order. After three sessions, the correlation between the mean errors found in these sessions was calculated. When the correlation coefficient was significant ($r > 0.78$, $p < 0.05$), subjects were asked to return to the lab 1 month later to perform a fourth measurement session. When these criteria were not met, stability of the patterns on a short time scale could not be proven and therefore stability of the patterns on a longer time scale was very unlikely. Therefore, subjects showing patterns that did not meet the criteria were not asked to perform a fourth session.

In experiment 2, subjects performed one experimental session of 1 h. Each force direction was presented ten times in a random order. In addition to the perception measurements, general subject parameters were recorded, which were: age, height, length of upper and lower arm, arm span, hand size, and Maximum Voluntary Contraction (MVC). MVC is a measure of hand strength.

In both experiments, subjects performed three practice trials to familiarize with the procedure.

3.2.4 Statistics

To assess whether subjects were veridical in their perception of force direction, t-tests were performed on the results per physical force direction for each subject, which showed whether the mean errors differed from 0. The influence of physical force direction on the error that subjects made in judging that direction was investigated using a repeated measured ANOVA with physical force direction as within-subject factor. In all ANOVAs, Greenhouse-Geisser correction was used when the sphericity criterion was not met. The similarities between and within subjects were assessed in more detail by calculating the correlation between the error patterns with Pearson's correlation tests. When the correlation coefficient was significant, it was deemed 'high'.

For experiment 1, the main effect of session on the error patterns was also assessed in the repeated measures ANOVA, which was performed on the data of the first three sessions only. To obtain a more detailed view of the similarities between the different sessions of single subjects, the correlations between the error patterns of the different sessions were calculated for each subject.

For experiment 2, the correlation of the error patterns between the subjects was tested. The significance of the general subject parameters as a predictor of the differences between subjects was measured by introducing these parameters as covariates in the repeated measures ANOVA and then testing their significance.

3.3   Results

A general overview of mean errors per physical force direction of both experiments can be found in Fig. 3.2. Note that the errors are quite large, even up to 60°. Moreover, the figure suggests that the patterns differ between subjects.

In the first experiment, 41 out of the 48 force direction-subject combinations had a mean that differed significantly from 0, which corresponds to 85% of the combinations. A significant effect of physical force direction on the error in perceived force direction was found ($F_{1.5,10} = 7.8$, $p = 0.012$). No main effect of

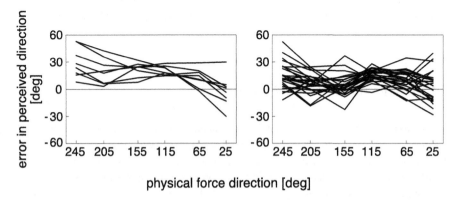

Fig. 3.2 Mean errors per physical force direction, as found in experiment 1 (left) and 2 (right). The means are based on 30 trials (3 sessions) per physical direction for experiment 1 and on 10 trials (1 session) per physical direction for experiment 2. The lines connect the errors per subject

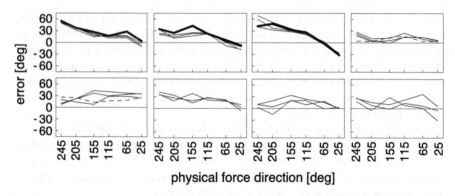

Fig. 3.3 Mean errors per physical force direction per subject, arranged from left to right in order of descending correlation between sessions. Thick solid lines indicate a fourth session with a high correlation with the first three sessions, while thin dashed lines indicate a low correlation. When only thin solid lines are plotted, subjects performed only three sessions

session on the error patterns was found ($F_{2,35} = 0.24$, $p = 0.79$). In Fig. 3.3, the error patterns at different sessions for all eight subjects are shown. In the first three sessions, six of the eight subjects showed high correlations between the sessions. For practical reasons, only five subjects performed a fourth session. In the fourth session, three of the five subjects showed an error pattern that was correlated to the patterns measured in the first three sessions. Overall, the correlation analysis revealed that 50% of the correlations between the first three sessions was high.

In the second experiment, 79 out of the 150 force direction-subject combinations had a mean that differed significantly from 0, which corresponds to 53% of the combinations. A significant effect of physical force direction on the error in perceived force direction was found ($F_{2.7,65} = 4.5$, $p = 0.0079$). The between-subject correlation revealed that only 7% of the error patterns showed a high correlation with patterns of other subjects. Figure 3.2 also visualizes the differences between subjects. None of the general subject parameters were significant when used as a covariate (all $F_{2.4,40} \leq 1.9$, all $p \geq 0.15$).

3.4   Discussion and Conclusion

In both experiments, we found that humans make substantial errors in judging the direction of a force, as more than half of the mean errors were non-zero. Moreover, these errors were different for different physical force directions, as shown by the significant effect of physical force direction on error. These results are congruent with our previous results [8], in which we found errors of similar magnitudes (also up to 60°).

In the previous study [8], we assessed perception of direction using forces of 2, 3, 4, 5 and 6 N. We found no influence of force magnitude on the error patterns.

Therefore, we chose to use only a force magnitude of 4 N in the current study. This is not a very large force magnitude, but it is far above the threshold of 0.1 N for feeling the difference between a right- and leftward force direction [2].

The results of experiment 1, based on 8 subjects, show that for half of the sessions, the patterns correlated within subjects over time. This suggests that the patterns are caused by internal subject parameters, rather than external ones that probably vary over time. For some subjects the patterns did not correlate so well over time, so external parameters might play some role, but internal parameters seem to be the most important ones.

The results of experiment 2, based on 25 subjects, show that only 8% of the error patterns were significantly correlated to patterns of other subjects. We first hypothesized that there might be groups of subjects with similar error patterns, but because of the poor correlation between subjects this does not seem likely. We then hypothesized that there is a general subject parameter that differs between subjects and that might explain these differences in error patterns. The absence of significance of any of the measured parameters (which were: age, height, length of upper and lower arm, arm span, hand size, and MVC) as a covariate in the model indicates that these parameters are not the ones that explain these differences.

Concluding, the correlation between the patterns within subjects (exp. 1) was much higher than the correlation between subjects (exp. 2). This seems to suggest that an internal parameter is the most important governing parameter for the errors in haptic perception of force direction. What this parameter is, remains to be investigated.

The knowledge that humans make errors in perceiving force direction should be considered in the design of force feedback algorithms. An application could, for instance, be the user-specific adjustment of force direction.

References

1. Barbagli F, Salisbury K, Ho C, Spence C, Tan HZ (2006) Haptic discrimination of force direction and the influence of visual information. ACM Trans Appl Percept 3(2):125–135
2. Baud-Bovy G, Gatti E (2010) Hand-held object force direction identification thresholds at rest and during movement. In: Kappers AML, Van Erp JBF, Bergmann Tiest WM, Van der Helm FCT (eds) Haptics: generating and perceiving tangible sensations. Lecture notes in computer science, vol 6192. Springer, Berlin/Heidelberg, pp 231–236
3. Coren S (1993) The left-hander syndrome. Vintage Books, New York
4. Elhajj I, Weerasinghe H, Dika A, Hansen R (2006) Human perception of haptic force direction. In: Proceedings of the IEEE/RSJ international conference on intelligent robots and systems, pp 989–993
5. Ho C, Tan HZ, Barbagli F, Salisbury K, Spence C (2006) Isotropy and visual modulation of haptic force direction discrimination on the human finger. In: Proceedings of eurohaptics 2006, pp 483–486
6. Tan HZ, Barbagli F, Salisbury K, Ho C, Spence C (2006) Force-direction discrimination is not influenced by reference force direction. Haptics-e 4:1–6

7. Toffin D, McIntyre J, Droulez J, Kemeny A, Berthoz A (2003) Perception and reproduction of force direction in the horizontal plane. J Neurophys 90(5):3040–3053
8. Van Beek FE, Bergmann Tiest WM, Kappers AML (2013) Anisotropy in the haptic perception of force direction and magnitude. IEEE Trans Haptics 6(4):399–407
9. Vitello MP, Ernst MO, Fritschi M (2006) An instance of tactile suppression: active exploration impairs tactile sensitivity for the direction of lateral movement. In: Proceedings of eurohaptics 2006, pp 351–355
10. Yang XD, Bischof W, Boulanger P (2008) The effects of hand motion on haptic perception of force direction. In: Ferre M (ed) Haptics: perception, devices and scenarios. Lecture notes in computer science, vol 5024. Springer, Berlin/Heidelberg, pp 355–360

Chapter 4
Perception of Force Magnitude and Postural Arm Dynamics

Abstract In a previous study, we found the perception of force magnitude to be anisotropic in the horizontal plane. In the current study, we investigated this anisotropy in three dimensional space. In addition, we tested our previous hypothesis that the perceptual anisotropy was directly related to anisotropies in arm dynamics. In experiment 1, static force magnitude perception was studied using a free magnitude estimation paradigm. This experiment revealed a significant and consistent anisotropy in force magnitude perception, with forces exerted along the line between hand and shoulder being perceived as 50% smaller than forces exerted perpendicular to this line. In experiment 2, postural arm dynamics were measured using stochastic position perturbations exerted by a haptic device and quantified through system identification. By fitting a mass-damper-spring model to the data, the stiffness, damping and inertia parameters could be characterized in all the directions in which perception was also measured. These results show that none of the arm dynamics parameters were oriented either exactly perpendicular or parallel to the perceptual anisotropy. This means that endpoint stiffness, damping or inertia alone cannot explain the consistent anisotropy in force magnitude perception.

Previously published as:
F.E. van Beek, W.M. Bergmann Tiest, W. Mugge & A.M.L Kappers (2015)
Haptic perception of force magnitude and its relation to postural arm dynamics in 3D
Scientific Reports, 5:18004.

4.1 Introduction

The use of teleoperation systems in dangerous, remote and small-scale environments is increasing. To provide the operator with a more complete picture of the environment at the slave side, providing haptic feedback in addition to visual feedback can be useful [1]. To understand how to design force feedback that is intuitive for the user, knowledge on human force perception is important. Our previous study [42] has shown that the perception of force magnitude in 2D (in the horizontal plane) is significantly and systematically anisotropic. This anisotropy fitted quite well with arm dynamics data from literature [21, 41]. Nonetheless, important questions still remained unanswered. How is force magnitude perception distributed outside of the horizontal plane? And is this distribution indeed related to arm dynamics? Force perception in 3D will be anisotropic, given that it is anisotropic in the horizontal plane. However, the orientation and shape of the complete anisotropy cannot be predicted from the 2D data. The direction in which forces are perceived as being the largest could, for instance, already have been found in the 2D measurements in the horizontal plane, or it could be more vertically oriented. Furthermore, the shape of the 3D anisotropy could be very elongated or more spherical. From a fundamental point of view, it is interesting to obtain a complete picture of the orientation and shape of the anisotropy in force magnitude perception, as the perception of forces is crucial in the interaction of humans with their environment. Moreover, it is still unknown if there is a link between arm dynamics and force magnitude perception [8]. So, our aim is to determine if this link indeed exists and therefore if the anisotropy in arm impedance might explain the perceptual anisotropy, by assessing if both data sets are oriented perpendicularly. Our hypothesis is that the direction in which forces are perceived as being the largest is perpendicular to the direction in which arm dynamics parameters are the largest, as the latter is the direction in which it is easier to resist the force. To this end, a good description of the shapes of both data sets in 3D is needed. So, in summary, the previous study was extended in two ways: (1) perception was measured in three dimensions and (2) postural arm dynamics were measured to be able to directly relate arm dynamics and force perception.

Force magnitude perception in directions other than that of gravity has mainly been studied in experiments investigating discrimination thresholds [12, 32, 45, 46]. In the current article, we focus on the effect of direction on the perceived magnitude of the force, which has only been studied by two research groups [12, 39]. Dorjgotov et al. [12] asked participants to discriminate between the magnitude of a reference force, exerted along the dorso-ventral axis towards the participant, and a test force exerted along one of the other cardinal axes. Forces were exerted through a haptic device with a ball-shaped gimbal, which participants enclosed with the whole hand. For all the test axes, a significant perceptual difference with the reference axis was found, indicating that a force towards the participant was perceived as being smaller than a force exerted along any of the test axes. The bias between test and reference axis was comparable for all test axes, which suggests that there was no perceptual difference between the test axes. However, the test axes were not compared directly and no other directions than the cardinal axes were tested. Tanaka and Tsuji [39] found elliptical anisotropies in a study using large forces in a similar task: a comparison between a tangential reference force and test forces along eight directions in the horizontal plane. A more direct way to investigate the distribution of force magnitude perception is by using a free magnitude estimation paradigm, which determines the relation between physical and perceived magnitude [18]. We used this method in our previous study [42] to assess force magnitude perception in several non-cardinal directions in the horizontal plane, in which we found an elliptically shaped distortion. In the present study, our main aim is to investigate the shape of the perceptual anisotropy in 3D, by extending our perceptual measurements to directions outside of the horizontal plane. To assess if anisotropies in arm dynamics are directly related to the shape of the perceptual anisotropy and could therefore be an explanation for it, we also measured arm dynamics.

Research into arm dynamics started by using step-like position perturbations to measure the stiffness, damping and inertia of the arm [2, 3, 11, 15, 27, 37, 41]. Another approach, which has gained influence over the last years, is the use of stochastic perturbations to determine arm impedance [9, 28]. In this approach, force (or position) is imposed as a continuous and unpredictable perturbation on the arm, while position (or force) is measured. The transfer function between the position and force signals describes arm admittance (or impedance), while the only prior assumption is linearity. Afterwards, models can be fitted to the measured transfer function to describe it (see e.g.[43]); a mass-damper-spring model seems to be a fair approximation for small arm displacements, when participants are asked not to intervene with the perturbations [31]. In several studies, correlations between arm dynamics and motor task execution have been found (e.g.[4, 6, 14, 21, 35, 40]). For instance, Trumbower et al. [40] showed that when participants are free to choose a posture in a tracking task in an unstable environment, they select a posture that shifts maximum arm impedance towards the direction of instability. Sabes et al. [35] describe that humans tend to adjust movements around obstacles by rotating the inertia ellipsoid in such a way that the inertia of the arm is maximal at the point where collision with the object is most likely. Cos et al. [6] showed that humans choose a biomechanically optimal path to a target, when choosing between two paths that do not differ in any other aspect. Most strikingly, humans are able to make this decision within 200 ms. A link between arm impedance and motor task execution is not surprising, as motor task execution is obviously influenced by arm dynamics. By measuring arm dynamics and comparing them to perceptual data, we investigate the presence of such a link between arm dynamics and a perceptual task.

The main aim of this study was to investigate the shape of the anisotropy in force magnitude perception in 3D. Furthermore, we measured postural arm dynamics – the impedance of the arm in a given posture – to investigate its correlation with the perceptual data. In experiment 1, force perception was measured using a free magnitude estimation paradigm. In experiment 2, postural arm dynamics were measured using a variation of a method described previously, using stochastic position perturbations [21, 22, 40]. Our data provide information about anisotropies in human force perception and arm dynamics and test the hypothesis that anisotropies in force magnitude perception are directly related to anisotropies in arm dynamics. If force perception could be predicted by modeling arm dynamics, predictions concerning, for instance, the effect of different postures could be verified. So, the application of this fundamental knowledge could contribute to designing force-feedback algorithms for teleoperation systems that correct for anisotropies in human force perception.

4.2 Methods

4.2.1 Participants

In both experiments, twelve right-handed (assessed using the Coren-test for handedness [5]) participants took part. In experiment 1, 4 males and 8 females participated, aged 23 ± 3 years (mean $\pm$ standard deviation), height 1.79 ± 0.09 m, with no known neurological disorders. In experiment 2, a new group of 12 right-handed people participated, consisting of 4 males and 8 females, aged 29 ± 3 years, height 1.73 ± 0.09 m. All participants gave written informed consent to participate in the study and were naive to the purpose of the experiment. For experiment 1, they received a compensation of 8 euros per hour. Prior to the experiments, they were given written and oral instructions on how to perform the experiment. From each participant, the following measures were recorded: height, arm span, shoulder-to-shoulder length, upper arm length and lower arm length. Both experiments were approved by the Ethics Committee of the Faculty of Human Movement Sciences (ECB) and were carried out in accordance with the approved guidelines.

4.2.2 Setup

Figure 4.1 shows an overview of the experimental setup. Participants were seated on a height-adjustable chair. In experiment 1, they were blindfolded. In experiment 2, a black cloth attached to the frame in front of their head prevented them from seeing their hand, while they could still see the screen, which was used for visual feedback. Their torso was fixed to the back of the chair using shoulder straps. Participants were asked to hold the handle of a haptic device, the HapticMaster (Moog Inc.) [44]. To prevent a change of posture, elbow height was fixed using an elbow sling, which was attached to the frame. The height and position of the chair, the position of the frame and the position of the elbow sling were adjusted to ensure that all participants performed the experiment using the same posture, with a slightly flexed, abducted shoulder and a slightly flexed elbow. Webcams were placed above and at the right side of the setup to check the angles of the upper and lower arm. For all participants, the following angles were used: torso-upper arm 38° in side view (see Fig. 4.1a for this angle) and 98° in top view, upper arm-lower arm 125° in top view (see Fig. 4.1b for the latter two angles). The lower arm was positioned in the horizontal plane, so the height of the arm sling was the same as the height of the handle of the haptic device, whereas the central position of the handle of the device was always in front of the sternum of the participant.

During experiment 2, force and position data were recorded by the HapticMaster at a sampling rate of 256 Hz. The recorded force was the force exerted on the ball-shaped handle (diameter 42 mm) of the haptic device. The recorded position was the position of the center of the handle of the device, which was rigidly

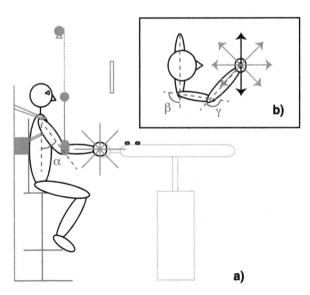

Fig. 4.1 Schematic representation of the setup and the posture of the participants. The joint angles were the same for all participants. (**a**) Side view of the setup. In dark gray, the back and top frame are drawn. The back frame was attached to the back of the chair, while the top frame was positioned above the participant. The back frame prevented movement of the chair, while the shoulder straps kept the participant in the same position throughout the experiment. A sling was attached to the top frame, in which the lower arm was inserted, to make sure that only small movements of the elbow in the horizontal plane were possible. Above the top frame, a webcam was mounted to check the posture of the participants. In gray, the chair is shown, which could be adjusted in height and position before it was attached to the back frame. The HapticMaster is show in gray, and the infrared Optotrak markers on top are drawn in black filled with light gray, while the participant is shown in black. In dark gray, the four planes in which the 26 force directions were defined are shown. In experiment 1, participants were blindfolded. In experiment 2, participants received visual feedback via the screen (open rectangle), which they could see through a gap in the frame (dashed line). A cloth (underneath frame) prevented them from seeing their own hand. (**b**) Top view of the setup, showing the 8 force directions that were defined in each of the planes. The black directions are the ones that are present in every plane, resulting in $4 \times 6 + 2 = 26$ unique directions. The dotted lines and arcs show the postural angles: $\alpha = 38°$, $\beta = 98°$, $\gamma = 125°$

connected to the device with a metal bar. During experiment 2, positions were also recorded using two infrared Optotrak markers, positioned on the horizontal tube of the HapticMaster (see Fig. 4.1a). These data were also collected with a frequency of 256 Hz. For the analysis, the position data measured with the Optotrak markers were used, as they did record position at endpoint level and therefore most closely resembled the true position of the handle. A comparison of the position data of the HapticMaster and those of the Optotrak revealed that both systems yielded the same results for frequencies below 10 Hz, which were the frequencies of interest for this experiment. However, our method was based on the method described by Trumbower et al. [40], in which Optotrak position data were used. For consistency, the same was done in our experiment.

4.2.3 Procedure

4.2.3.1 Experiment 1: Perception

During experiment 1, participants were presented with a force (2, 3, 4, 5, or 6 N) in one of the four planes shown in Fig. 4.1a. In each plane forces were presented in eight directions, as shown in Fig. 4.1b. This resulted in 26 unique predefined directions, in which forces were exerted on the participant's right hand using the handle of a HapticMaster. The force was increased from zero to the test force in 2 s using a linear force ramp, which thus had a different slope for every force level. Participants could therefore not use the duration of the force ramp as a cue to estimate the force level. When the force reached the test level, a tone indicated that participants could verbally indicate the magnitude of the force. They could use a number on any scale, as long as they kept their scale linear and used the same scale throughout each session. There was no time limit, but participants generally answered within a few seconds after the tone. When an answer was given, the force was ramped down in 1.5 s to zero again. After this, a relax phase of 3 s followed, after which the next trial started. During the force presentation, participants were instructed to keep their hand steady. If they moved their hand more than 3 cm from the starting point in any direction, a tone indicated they had exceeded the limits. If this happened, the trial was rejected and the handle moved back to the start position to initiate a new trial. All rejected trials were repeated at the end.

Experiment 1 consisted of 650 trials per participant, divided over 3 one-hour sessions. All direction-magnitude combinations were repeated 5 times. The order of combinations was pseudo-randomized, by making sure that participants first experienced all possible combinations once in a random order, after which all possible combinations were presented a second time in random order, etc. To avoid fatigue, each session was divided into 4 blocks of 52–55 trials, which took about 10 min per block. In between the blocks, participants were asked to relax their arm and hand and take some rest to make sure no effect of fatigue was present. They indicated when they were ready to proceed to the next block themselves. Participants were given some practice trials before starting the measurements. In these practice trials, they were given no feedback on their performance other than technical instructions on how to perform the task as it was intended.

4.2.3.2 Experiment 2: Arm Dynamics

In experiment 2, arm dynamics were measured in the 26 directions in which force perception was measured in experiment 1. Each trial started similar to those in experiment 1: a force of 4 N was ramped up linearly and participants were instructed to resist this force, resulting in a steady hand position. Eight seconds after reaching the plateau level of the constant force, 35 s of stochastic position perturbations started (see below for a description of the signal design). Participants were instructed to keep exerting the constant force of 4 N (which was always in the same direction as the perturbations), while not intervening with the perturbations in any way. By asking participants to keep exerting the force, the voluntary force level in experiment 2 was kept as comparable to that in experiment 1. The participants were aided in this task by feedback on a screen that showed the force target, which was the force ramp followed by the force plateau of 4 N. On top of that, the filtered measured force along the perturbation direction was plotted, using a running average filter with a window size of 94 ms. It was stressed to the participants that their task was only to keep the mean force at 4 N and not to intervene with fluctuations in the force signal around this mean force, which were induced by the perturbations.

Experiment 2 consisted of 26 trials, presented in a random order. Each perturbation signal lasted 35 s, resulting in a total time of ~50 s per trial. Between trials, participants could relax for a few seconds. There was a large break to avoid fatigue, dividing the experiment in two blocks of about 15 min each. Again, participants could indicate themselves when they felt ready to proceed to the next block. Participants were given some practice trials before starting the measurements. In these practice trials, they were given no feedback on their performance other than technical instructions on how to perform the task as it was intended.

4.2.4 Position Disturbance Signal Design

The position disturbance signals were designed as a one-dimensional version of the method described by Trumbower et al. [40]. For every trial, a white noise signal of 35 s was generated, that was filtered using a second-order Butterworth filter with a cut-off frequency of 4 Hz. The signal was scaled to an RMS amplitude of 5 mm, resulting in a maximum movement amplitude of about 20 mm. For an example of a section of such a signal, see Fig. 4.2a. The position signal was used as the input signal for the HapticMaster, which was programmed as a position perturbation system by using stiff (20 kN/m) springs, with a damping ratio of 0.3, to send the handle to the required positions.

For the arm dynamics measurements, a one-dimensional approach was taken, because it most closely resembled the situation in which perception was measured. For measuring perception, one force direction per trial was used, which is congruent to one-dimensional perturbations in the arm dynamics measurements. To make sure that the level and direction of voluntary force was the same during the arm dynamics measurements as during the perceptual measurements, the approach of asking participants to exert a bias force ('keep exerting this 4 N') on top of a relax task ('do not intervene with the perturbations') was necessary (see Mugge et al. [26] for a description of the influence of different task instructions on arm dynamics). We could have chosen to perform 3D perturbations for every bias force direction, which would have resulted in 26 slightly different ellipsoids per parameter and participant, which would all have been slight variations of the ellipsoids that could have been measured in a relax task [29]. However, we still would have only used one data point per ellipsoid in the analysis: the data point in the direction of the bias force. So, in

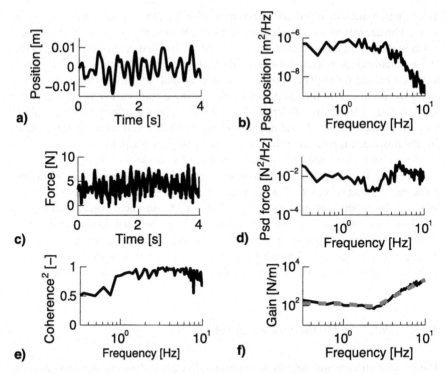

Fig. 4.2 Typical example of force and position data of one trial of one participant in the arm dynamics measurements. (**a–b**) Time and frequency domain representation of the position input signal, as measured using Optotrak markers at the end effector of the haptic device. Only a few seconds of the time domain representations are shown. (**c–d**) Time and frequency domain representation of the force signal, as measured with the force sensor of the HapticMaster. (**e**) Squared coherence of the position and force data. (**f**) Frequency Response Function based on the ratio of the spectra in **b** and **d**. The dashed gray line shows the fit that was made using the mass-damper-spring model (see Eq. 4.4). For this trial, the model yielded the following values: $M = 0.554\,\text{kg}$, $B = 6.45\,\text{Ns/m}$, $K = 135\,\text{N/m}$

our one-dimensional approach, we did not measure 26 data points on 1 ellipsoid, but gathered data of one specific direction of 26 ellipsoids per participant, which was sufficient to answer our research question.

4.2.5 Data Analysis

4.2.5.1 Perceptual Data

The mean of the magnitude perception data per session was scaled to the actual mean force of that session, which was always very close to 4 N (the precise value depended on the order of the magnitude-direction combinations). This was done

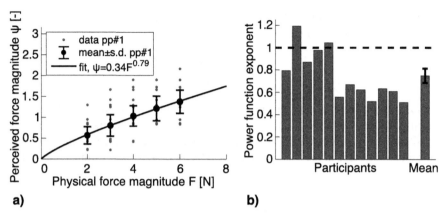

Fig. 4.3 Overview of the perceptual data, showing the relation between physical and perceived force magnitude, measured using free magnitude estimation. (**a**) Typical example of force magnitude perception data of one participant, for all directions together. The small grey dots show individual measurements, while the thick black dots with error bars show the mean per physical force magnitude ± standard deviation. The black line shows the fit of the power function, which describes the relation between physical force magnitude (F) and mean perceived force magnitude (ψ). The fit was used to normalize the data. (**b**) Exponents of the fitted power functions, per participant and mean ± standard error over participants. Note that on average, the power function exponents differ from 1, indicating a non-linear relationship between physical and perceived force magnitude

to correct for the fact that participants might have changed their scale between sessions. A power function was fitted to describe the relation between physical force magnitude and perceived force magnitude for the complete data set, as shown in Fig. 4.3. Subsequently, the mean perceived force magnitude values per physical force magnitude were calculated according to the fit. By dividing the actually perceived force magnitude values by these calculated mean values, the data were normalized. The normalized data were used in the further statistical analyses.

4.2.5.2   Arm Dynamics Data

Since position perturbations were used to measure arm impedance, an open-loop approach could be used for the analysis of the arm dynamics data (as shown in [9, 28]). For each trial, the registered force and position data (for examples, see Fig. 4.2a, b) were analyzed in the frequency domain, by first calculating the spectra of both signals using a Fast Fourier Transform (for examples, see Fig. 4.2c, d), from which cross-spectral densities could by determined. The first data point (at 0 Hz) was removed in all spectra to eliminate the constant force from the analysis. To obtain smoother cross-spectral densities and to obtain a better estimate of the coherence of the signals, averaging over four adjacent frequencies was used [17]. From these averaged cross-spectral densities, transfer functions (also called

Frequency Response Functions or FRFs) and their coherence were calculated. The transfer functions were calculated in this way:

$$H_{xf} = \frac{G_{xf}}{G_{xx}} \qquad (4.1)$$

with H_{xf} being the FRF for that trial, as illustrated in Fig. 4.2f, G_{xf} being the cross-spectral density of the position and force signal and G_{xx} being the power spectral density of the position signal. To assess the linearity of this response function and the amount of noise in the system, coherence (see Fig. 4.2e) was estimated using:

$$\gamma_{xf}^2 = \frac{|G_{xf}|^2}{G_{xx}G_{ff}} \qquad (4.2)$$

For a more elaborate description of the derivation of Eqs. 4.1, 4.2 and the averaging method, see De Vlugt et al. [9].

4.2.5.3   Impedance Model Fit

One of the most simple ways to model the human arm is by describing it as a mass-damper-spring system, of which the general formula in the time domain is:

$$M\ddot{x}(t) + B\dot{x}(t) + Kx(t) = f(t) \qquad (4.3)$$

with M, B and K being the inertia, damping and stiffness parameters describing the system, f being the force and x being the position. In the frequency domain, this function can be written as:

$$Ms^2 + Bs + K = H_{xf}(s) \qquad (4.4)$$

in which H_{xf} is the FRF, $s = 2\pi f\sqrt{-1}$ and f is the frequency. This model has proven to be a representative model of a human arm for small displacements, when participants are asked not to intervene with the perturbations, such as in the present study [11, 41]. Since we estimated the FRFs from our measured data, as shown in Eq. 4.1, we could fit the model in Eq. 4.4 to these FRFs to obtain the system parameters. The model was only fitted for frequencies below 10 Hz, since visual inspection of the FRFs and a decline in the coherences showed that the system started deviating from a second-order system above this frequency. Probably, contact dynamics started playing a role above 10 Hz [26], in which we were not interested. In these types of experiments, fits are commonly restricted to frequencies below 10 Hz [22, 29, 40]. The fitting was performed using a logarithmic least squares fitting procedure in which the following error was minimized:

$$\text{error} = \sqrt{\frac{\gamma_{xf}^2}{f+1}} \log \left| \frac{H_{xf_{\text{estimated}}}}{H_{xf_{\text{model}}}} \right| \tag{4.5}$$

with $H_{xf_{\text{estimated}}}$ being the FRF estimated from the measured data and $H_{xf_{\text{model}}}$ being the FRF resulting from the model in Eq. 4.4, when using the fit parameters M, B and K. The coherence of the measured signal (γ_{xf}^2) and the frequency (f) were used as weight factors, to make sure that low-coherent data points receive less weight and to compensate for the fact that at higher frequencies the data points are more tightly spaced in a logarithmic representation. An example of such a fit is shown in Fig. 4.2f. By performing this fit for all the FRFs in the 26 measured directions, 26 values for stiffness, damping and inertia were found per participant.

4.2.5.4 Statistics

To assess the effect of direction, a repeated measures ANOVA with direction as the within-subject variable was performed on the parameter perceived force magnitude for experiment 1 and on the parameters stiffness, damping, and inertia for experiment 2. For the analysis of experiment 1, the interaction between the effect of force direction and the variables physical force and session was also tested. When the sphericity criterion was violated, Greenhouse Geisser correction was used.

To obtain a description of the orientation of the data, a principal component analysis (PCA) was performed on the 26 values found for perceived force magnitude, stiffness, damping, and inertia, for each participant. This analysis describes data using three orthogonal axes for three dimensional data, by aligning the axes with the directions with the largest amount of variance in the data. Therefore, the first component of the PCA shows the major axis of the data, while the third component reveals the minor axis. From these components, the eccentricity was calculated, which is the ratio between the minor and major axis. The similarity of the orientations of the perceptual data set and the data sets of each of the arm dynamics parameter was assessed by testing the similarity in major axis direction. To obtain one value per vector describing the orientation, all vectors of each combination of perceptual data and one of the arm dynamics parameters were projected on a 2D plane. This 2D plane was defined as the plane spanned by the mean vector of the perceptual data and the mean vector of the arm dynamics parameter of interest.

When a test was needed to assess if the values in a data set differed from 1, a one-sample t-test was used. When the means of two independent data sets had to be compared, an independent samples t-test was used, with Bonferroni correction when appropriate. When equal variances could not be assumed, the degrees of freedom and p-value were adjusted accordingly. For all statistical analyses, $p < 0.05$ was deemed significant.

4.3 Results

4.3.1 Experiment 1: Perception

In experiment 1, participants verbally reported the perceived magnitude of force stimuli, which could be exerted in 26 directions and with 5 different magnitudes. A typical example of force perception data is shown in Fig. 4.3a. For each physical force magnitude, some spread in the perceived force magnitudes can be seen (the spread of the thin gray dots around the thick black dots in Fig. 4.3a), which had a standard deviation of 0.30 on average across participants and physical force magnitudes. Power functions were fitted to the perceptual data of each participant separately, averaged per physical force magnitude. Since a power function can only describe the mean variation caused by the independent variable and not the noise around the mean [18], the fit was performed on the mean values per physical force magnitude. An example of such a fit is also shown in Fig. 4.3a. The goodness-of-fit was excellent, with R^2 values being greater than 0.98 for each participant.

In Fig. 4.3b, the fitted power function exponents are shown for each participant. For 10 of the 12 participants, the exponent was lower than 1, suggesting a non-linear relationship between physical and perceived force magnitude. A one-sample t-test on the exponents showed that they indeed differed significantly from 1 ($t_{11} = -3.9$, $p = 0.003$). Since the relation between physical and perceived force magnitude was non-linear, normalizing the data across physical force magnitudes could not be done by dividing the perceived magnitude by the physical magnitude. To still be able to assess the influence of direction, irrespective of physical force magnitude, normalization was done by calculating the fitted perceived force magnitude values per physical force magnitude according to the fitted power function. The actually perceived force magnitude values were divided by these calculated fit values to normalize the data.

On these normalized data, a repeated measures ANOVA was performed, which showed a significant effect of force direction on perceived force magnitude ($F_{25,275} = 10.6$, $p < 0.001$). This means that force magnitude perception is indeed anisotropic in 3D. To assess if the pattern of the perceptual anisotropy was different between the three measurement sessions or between the five physical force magnitudes, the interactions between session and force direction and between physical force magnitude and force direction were also tested in a repeated measures ANOVA. These interactions were not significant ($F_{1.1,60} = 1.31$, $p = 0.27$ and $F_{7.3,80} = 1.43$, $p = 0.20$, respectively), indicating that the patterns did not differ between sessions and physical force magnitudes. This shows, in retrospect, that averaging over sessions and physical force magnitudes did not influence the results.

A Principal Component Analysis (PCA) was performed to assess the orientation and the shape of the perceptual data set. An overview of the principal components is given in Fig. 4.4, which shows a striking similarity across participants and physical force magnitudes. The analyses showed that the ratio between the smallest and the largest value (i.e. the ratio between the eigenvalues of the third and first component

**Components per participant
(n=12)**   **Components per force magnitude
(2-6 N)**

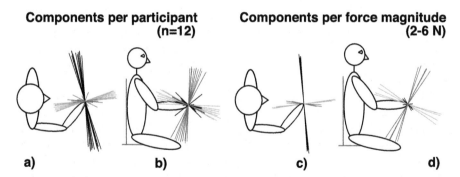

a)   b)   c)   d)

Fig. 4.4 Perceptual data, showing the three axes of the Principal Component Analysis (PCA) per participant and per physical force magnitude. Black: first component, dark gray: second component, light gray: third component. (**a**) Top and (**b**) side view, with each vector corresponding to one participant. (**c**) Top and (**d**) side view, with each vector corresponding to one physical force magnitude, while the data were averaged over participants. The first component is considerably larger than the second and third component, resulting in eccentricities (i.e. the ratio between the eigenvalues of the third and first component of the PCA) that were well below 1. The orientation and size of the components are consistent over participants and physical force magnitudes. Together, this illustrates the consistent anisotropy in force magnitude perception

of the PCA), which we will call eccentricity, was 0.66 ± 0.02 (mean $\pm$ standard error). The similarity of the orientation of the principal components and the eccentricity being considerably smaller than 1 together indicate a systematic distortion in force magnitude perception.

4.3.2 Experiment 2: Arm Dynamics

In experiment 2, postural arm dynamics data were recorded at end-point level along the same 26 directions that were used in experiment 1. In Fig. 4.2, example data of one trial and all steps in the analysis of these data are shown. The analysis yielded values for arm stiffness, damping and inertia, for each participant and each direction. The consistency of the fit values across participants was determined by calculating the standard error of the values across participants per direction, after which these values were averaged across directions, resulting in an indication of the mean variation across participants per parameter. These calculations yielded a standard error of 22% for stiffness, 14% for damping and 13% for inertia. The average R^2 over all participants and directions was 0.77 ± 0.03 (mean $\pm$ standard error). The analytical steps are described in more detail in the 'Methods' section. A repeated measures ANOVA on the stiffness, damping and inertia values showed that there was a significant effect of direction on all parameters (all $F_{25,275} > 10.6$, all $p < 0.001$). A PCA was performed on all three arm dynamics parameters for each participant, to reveal the orientation and shape of the data sets. These analyses yielded eccentricities of 0.73 ± 0.02 for stiffness, 0.51 ± 0.02 for damping, and 0.56 ± 0.02 for inertia (mean $\pm$ standard error). Clearly, all eccentricities are considerably smaller than 1, as the mean values are all more than 10 standard errors below 1. This confirms the effect of direction on all arm dynamics parameters, which means that there is an anisotropy in the arm dynamics parameters.

The next question is if the anisotropies in arm dynamics data are related to the anisotropy in the perceptual data, i.e. if the perceptual data set is oriented perpendicular or parallel to one of the arm dynamics data sets. Experiment 1 and 2 were performed with different participants, so a comparison between the vectors on participant level was not possible. Instead, the perceptual vectors and the arm dynamics vectors were averaged to compare their mean orientation. To statistically compare the data sets, the orientation of each vector for each participant needed to be calculated, preferably using a single measure. To do this for each combination of perceptual data and one of the arm dynamics parameters, the vectors of both data sets were projected onto a plane spanned by the mean vectors of both data sets. So, to compare the perceptual and stiffness data, for instance, all stiffness and perception vectors were projected onto the plane spanned by the mean perception and the mean stiffness vector. The projections of all combinations of data sets are shown in Fig. 4.5a–c. For all combinations, the planes of projection were close to the plane spanned by the first and third component of the PCA on the perceptual data (see Fig. 4.4 for the PCA on the perceptual data). The angle between the projected data sets was $70 \pm 5°$ for perception and stiffness, $54 \pm 2°$ for perception and damping, and $39 \pm 4°$ for perception and inertia (mean $\pm$ standard error.). To assess if these angles differed significantly from 0 or $90°$, 2-sample t-tests with a predicted mean of respectively 0 and $90°$ were performed for each combination. If the data sets were oriented parallel ($0°$), this would mean that the direction of largest perception and arm dynamics values coincided, while a perpendicular orientation ($90°$) would mean that the direction of the smallest perceptual values coincided with the largest arm dynamics values. However, the angles between the projected arm dynamics and perceptual data were significantly different from $0°$ (perception-stiffness: $t_{14} = 15$, $p < 0.001$, perception-damping: $t_{22} = 28$, $p < 0.001$, perception-inertia $t_{22} = 11$, $p < 0.001$) and from $90°$ (perception-stiffness: $t_{14} = -4.4$, $p = 0.004$, perception-damping: $t_{22} = -19$, $p < 0.001$, perception-inertia $t_{22} = -14$, $p < 0.001$). An overview of the angles between the projected axes is given in Fig. 4.5d.

4.4   Discussion

In this study, we found systematic anisotropies in force magnitude perception and in arm dynamics in 3D. These anisotropies were consistent over participants for both experiments. However, the anisotropies in arm stiffness, damping and inertia were not oriented perpendicular or parallel to the perceptual anisotropies. The significant anisotropy in force magnitude perception in 3D is consistent with the anisotropy that we found in the horizontal plane in our previous study [42]. In three dimensions,

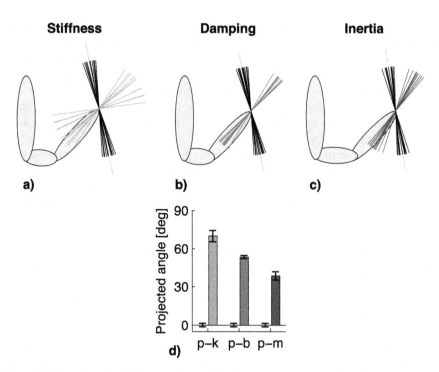

Fig. 4.5 Comparison of the 2D projections of the first component of the PCA of perceptual data and the first component of the PCA of the three arm dynamics parameters. Each subplot shows a projection onto a slightly different plane, as the plane is determined by the mean vector of the perceptual data and the mean vector of the arm dynamics parameter of interest. Each line shows the orientation of the first component of the PCA of the data of one participant. All vectors are unit vectors. The light gray shapes show the projection of shoulders, upper and lower arm of a participant in the experimental posture. (**a**) Perception (black) and stiffness (light gray) vectors. (**b**) Perception (black) and damping (gray) vectors. (**c**) Perception (black) and inertia (dark gray) vectors. (**d**) Projected angle of each of the data sets shown in panel **a–c**, with respect to the light gray axis (which is defined by the mean of the perceptual data set, which is obviously 0 in this definition). Note that none of the arm dynamics parameters are oriented perpendicular or parallel (i.e. 90° or 0°) to the perceptual data

the shape of this distortion is somewhat cigar-like, with its major axis oriented at a small angle with the transversal axis, while the minor axis is oriented approximately along the line between hand and shoulder (see Fig. 4.4). So, both the major and the minor axis lie close to the horizontal plane. On average, the perceptual eccentricity was 0.66, which means that a force exerted along the major axis was perceived as being 50% larger than when it would have been exerted along the minor axis (since $1/0.66 \approx 1.5$). The orientation and shape of the anisotropy was similar for all participants and for the five different force magnitudes that were used (2–6 N), which shows the consistency of the effect. Interestingly, our previous experiment [42] was performed with a different setup with a different handle, which was a tube-shaped handle that was grasped in a power grip, while the current setup had a ball-shaped handle. Apparently, this did not have a significant influence on the perceptual results. The knowledge that force perception is anisotropic could be used in, for instance, force-feedback devices. In these devices, the magnitude of force feedback could be adjusted according to the anisotropies, which would result in an isotropic perception of force. In everyday life, we apparently correct for the anisotropic nature of force perception in motor task execution, since we interact naturally with objects in many configurations. However, when force-feedback is used in a haptic guidance-type of feedback [1], the forces are not directly coupled to the environment anymore, but they are used to provide additional information. For instance, when haptic guidance along a desired movement trajectory is provided, it might be beneficial to provide users with perceptually consistent, rather than physically veridical force feedback, to make sure that the level of guidance is perceptually equal along different directions. Moreover, although humans may be able to correct for perceptual anisotropies, it does not necessarily mean that physically veridical feedback is optimal. In critical situations, conflicts between modalities might make a difference, for instance by increasing reaction times. An attempt to incorporate force perception anisotropies, inspired by fundamental work on perception of force and position, has been made in a haptic feedback device for kinesthetic training [8].

The measurements of arm dynamics also yielded consistent results over participants, shown by the consistency in anisotropy orientation and by the relatively small variation in absolute values across participants. The orientation of the anisotropies in stiffness, damping and inertia found in our study were comparable to anisotropies found for 3D measurements of postural arm dynamics [21, 31]. In the perceptual task, participants were asked to resist a force with a gradual ramp, which only resulted in limited movements around the start position, especially since trials with substantial movement were rejected. Since the movements were small, we compared our perceptual data to postural impedances. Our postural impedance values, measured in Experiment 2 using small perturbations around a given posture, were substantially higher than values found in the literature for tasks in which movement to a target position is required [7]. It is therefore likely that our measurements were dominated by contributions of short-range impedance [19, 34]. In specific perturbation directions the rotations may have been sufficiently large to exceed the domain of short-range stiffness. Nonetheless, our absolute arm dynamics values were comparable to values found in other studies measuring postural arm dynamics [13, 30].

The arm dynamics that we measured were likely composed of intrinsic muscle properties and reflexive feedback [7, 10, 13, 36]. These components were likely also present during the perceptual measurements. By using a one-dimensional approach for the arm dynamics measurements, we measured the anisotropies in a situation comparable to the perceptual measurements, as discussed in more detail in the 'Methods' section. To ensure that the contribution of all parameters was comparable in both experiments, we controlled the parameters that we regarded as the most important parameters influencing the orientation of arm dynamics: posture and task instruction. The same joint angles were used in both experiments, as posture is a very important determinant in the orientation of the stiffness, damping and inertia anisotropies [25, 41]. Our task instruction was also aimed at keeping the situation comparable in both experiments. Firstly, we instructed participants to exert a constant force, using feedback of their force on a screen. Voluntary force generation is known to influence particularly the stiffness parameter [29]. Secondly, we instructed participants to not intervene with the perturbations. This instruction was meant to minimize the contribution of voluntary control and reflexive feedback. In the next paragraphs, both aspects of the instruction will be discussed in more detail.

The instruction 'keep exerting this force' was used to keep the contribution of voluntary force comparable in both experiments, because it has been found that voluntary force influences particularly the stiffness parameter [20, 30]. Perreault et al. [29] have shown that the orientation and area of stiffness ellipses measured in the horizontal plane are affected by the level of voluntary force. However, even the lowest voluntary force that they used, around 12 N, was much higher than the voluntary force used in our current experiment (4 N), while the effects on ellipse orientation that they reported were most prominent at very high forces (up to 72 N). Dolan et al. [11] found no differences between the orientation of stiffness ellipses measured using a voluntary force of 1 N and those measured during a relax task. Furthermore, Krutky et al. [21] showed that when participants interacted with environments that were unstable along one of the cardinal axes, and therefore could have induced stiffness adaptations in these directions, there was no difference between the orientations of the stiffness ellipsoids measured in 3D in the different unstable environments. An increase in overall ellipsoid size was found for stiffness, but for damping and inertia no changes in ellipsoid orientation or size were reported. Even when participants are asked to actively change the orientation of their stiffness ellipse by using visual feedback, they are only able to do so to a small extent when they concurrently have to exert a voluntary force [30]. So, we believe that the main difference between arm dynamics parameters that we would have measured if we would have used a true relax task and the parameters that we measured in our current experiment is a difference in size and not in orientation of the parameters. Since we are only interested in the orientation of the parameters and not in the absolute values, we consider the addition of a voluntary force to the arm dynamics measurement to be a cautious addition.

The instruction 'do not intervene with the perturbations' was used to minimize the contribution of voluntary control and reflexive feedback. In experiment 1, intrinsic muscle properties and reflexive feedback undoubtedly both contributed to the total impedance of the arm [7, 10, 13, 36], so they were also measured together in experiment 2. EMG signals were not recorded, so we could not quantify the contribution of the reflexive component directly. It would have been an option to include reflexive feedback in our model of the system [26]. However, as reflected by the good fits with the mass-damper-spring model without reflexive feedback, the contribution of reflexive feedback was small in our particular task [26]. Apparently, the participants minimized responses to the position disturbances following their task instruction. In position tasks (e.g. [33]) or tasks involving movements to a target position (e.g. [23]) reflexive feedback does substantially contribute to arm impedance, but its contribution is small in tasks in which participants are asked not to intervene with the disturbances (e.g. [10, 36]). Forbes et al. [13] show that when participants are asked to increase their arm stiffness based on feedback of their EMG signals, they mainly increase their intrinsic stiffness, while hardly increasing their reflexive stiffness. In our experiment, we essentially do the same by asking participants to keep exerting a force (which corresponds to the required EMG level) and to not intervene with the perturbations (which corresponds to the small contribution of reflexive feedback). So, we used intrinsic parameters only to represent the human arm, because this is a simple and robust model, with minimal redundancy, while it is sufficient to answer our research question and captures the most important impedance properties. The adequate goodness-of-fit values and the consistency of the results across participants show that this model does provide a proper description of the measured arm impedance. Moreover, we again point out the small magnitude of the voluntary force in this experiment, making the task more closely resemble a true relax task, in which studies report minimal reflexive feedback, than a position task, in which studies do report considerable reflexive feedback [10, 36].

Our data indicate that stiffness, damping or inertia alone cannot explain the anisotropy in force magnitude perception, since none of the arm dynamics anisotropies were oriented perpendicular or parallel to the perceptual anisotropy, while all anisotropies were consistent over participants. However, this does not mean that there is no relation between force perception and arm dynamics. Because of the static nature of the perception task, we think that stiffness is the most likely candidate of the arm dynamics parameters to be involved. Stiffness was also the parameter that was the closest to perpendicular to the perceptual data set (70°). A perpendicular orientation of the data sets would have supported the intuitively logical explanation that large stiffnesses coincide with smaller perceived forces. However, the significant deviation from 90° shows that endpoint stiffness alone cannot be the parameter causing the consistent anisotropy in force magnitude perception. An alternative explanation could be sought in the simplicity of the model of the arm. One important assumption is that the whole arm acts as a single unit and no bi-articular muscles are involved, which is obviously a simplification. Bi-articular muscles cause joints to be coupled, which makes the prediction of stiffness at endpoint level more complex, because bi-articular muscles can exert force in different directions and with different magnitude, depending on which other muscles are activated concurrently [16, 24]. This coupling has been shown to cause a directionally selective increase in stiffness when making movements in unstable environments [14], although Perreault et al. [29] report that bi-articular muscles play a smaller role in the total endpoint stiffness than mono-articular muscles do during measurements of postural arm dynamics. Speculating on this topic, we suggest that if bi-articular muscles indeed played a role in our measurements, then this could have resulted in an irregular shape of the stiffness distribution. If the shape of the distribution would be irregular and would, for instance, have multiple 'major axes', the major axis determined with a PCA does not necessarily describe the direction of the largest values. This could mean that the actual orientations of the largest arm stiffness values were somewhat different than the ones found in our approach. Another aspect that might be important for the anisotropy in force perception is the contribution of different types of information to the total perceptual result, such as information from muscle spindles, tactual information and the efference copy. For instance, producing a force yields a smaller estimation of its magnitude than perceiving a force does, which is thought to be a result of the efference copy in force production [38]. What the effect of these speculations would be on the assessment of the relation between arm dynamics and force perception, remains to be investigated.

Concluding, we found that force magnitude perception is anisotropic in 3D. This anisotropy was not directly related to arm stiffness, damping or inertia, when modeling the human arm as a single unit. A more detailed model of the arm, possibly including bi-articular muscles, might be needed to understand the nature of the anisotropy in force magnitude perception.

References

1. Abbink DA, Mulder M, Boer ER (2012) Haptic shared control: smoothly shifting control authority? Cogn Tech Work 14(1):19–28
2. Artemiadis P, Katsiaris P, Liarokapis M, Kyriakopoulos K (2010) Human arm impedance: characterization and modeling in 3D space. In: International conference on intelligent robots and systems, pp 3103–3108
3. Artemiadis P, Katsiaris P, Liarokapis M, Kyriakopoulos K (2011) On the effect of human arm manipulability in 3D force tasks: towards force-controlled exoskeletons. In: IEEE international conference on robotics and automation, pp 3784–3789
4. Burdet E, Osu R, Franklin DW, Milner TE, Kawato M (2001) The central nervous system stabilizes unstable dynamics by learning optimal impedance. Nature 414(6862):446–449
5. Coren S (1993) The left-hander syndrome. Vintage Books, New York
6. Cos I, Duque J, Cisek P (2014) Rapid prediction of biomechanical costs during action decisions. J Neurol 112(6):1256–1266
7. Crevecoeur F, Scott SH (2014) Beyond muscles stiffness: importance of state-estimation to account for very fast motor corrections. PLoS Comput Biol 10(10):e1003869
8. De Santis D, Zenzeri J, Masia L, Squeri V, Morasso P (2014) Exploiting the link between action and perception: minimally assisted robotic training of the kinesthetic sense. In: 5th IEEE RAS EMBS international conference on biomedical robotics and biomechatronics, pp 287–292
9. De Vlugt E, Schouten AC, van der Helm FCT (2003) Closed-loop multivariable system identification for the characterization of the dynamic arm compliance using continuous force disturbances: a model study. J Neurosci Methods 122(2):123–140
10. De Vlugt E, Schouten AC, van der Helm FCT (2006) Quantification of intrinsic and reflexive properties during multijoint arm posture. J Neurosci Methods 155(2):328–349
11. Dolan J, Friedman M, Nagurka M (1993) Dynamic and loaded impedance components in the maintenance of human arm posture. IEEE Trans Syst Man Cybern 23(3):698–709
12. Dorjgotov E, Bertoline G, Arns L, Pizlo Z, Dunlop S (2008) Force amplitude perception in six orthogonal directions. In: Proceedings of the symposium on haptics interfaces for virtual environment and teleoperator systems, pp 121–127
13. Forbes PA, Happee R, van der Helm FCT, Schouten AC (2011) EMG feedback tasks reduce reflexive stiffness during force and position perturbations. Exp Brain Res 213(1):49–61

14. Franklin D, Burdet E, Osu R, Kawato M, Milner T (2003) Functional significance of stiffness in adaptation of multijoint arm movements to stable and unstable dynamics. Exp Brain Res 151(2):145–157
15. Gomi H, Kawato M (1997) Human arm stiffness and equilibrium-point trajectory during multi-joint movement. Biol Cybern 76(3):163–171
16. Hogan N (1985) The mechanics of multi-joint posture and movement control. Biol Cybern 52(5):315–331
17. Jenkins GM, Watts DG (1968) Spectral analysis and its applications. Holden-Day, San Francisco
18. Jones LA, Tan HZ (2013) Application of psychophysical techniques to haptic research. IEEE Trans Haptics 6(3):268–284
19. Joyce GC, Rack PMH, Westbury DR (1969) The mechanical properties of cat soleus muscle during controlled lengthening and shortening movements. J Physiol 204(2):461–474
20. Kearney RE, Hunter IW (1989) System identification of human joint dynamics. Crit Rev Biomed Eng 18(1):55–87
21. Krutky MA, Trumbower RD, Perreault EJ (2009) Effects of environmental instabilities on endpoint stiffness during the maintenance of human arm posture. In: Annual international conference of the IEEE Engineering in Medicine and Biology Society, pp 5938–5941
22. Krutky MA, Ravichandran VJ, Trumbower RD, Perreault EJ (2010) Interactions between limb and environmental mechanics influence stretch reflex sensitivity in the human arm. J Neurophysiol 103(1):429–440
23. Kurtzer I, Crevecoeur F, Scott SH (2014) Fast feedback control involves two independent processes utilizing knowledge of limb dynamics. J Neurophysiol 111(8):1631–1645
24. McIntyre J, Mussa-Ivaldi FA, Bizzi E (1996) The control of stable postures in the multijoint arm. Exp Brain Res 110(2):248–264
25. Milner TE (2002) Contribution of geometry and joint stiffness to mechanical stability of the human arm. Exp Brain Res 143(4):515–519
26. Mugge W, Abbink DA, Schouten AC, Dewald JPA, Helm FCT (2010) A rigorous model of reflex function indicates that position and force feedback are flexibly tuned to position and force tasks. Exp Brain Res 200(3–4):325–340
27. Mussa-Ivaldi F, Hogan N, Bizzi E (1985) Neural, mechanical, and geometric factors subserving arm posture in humans. J Neurosci 5(10):2732–2743
28. Perreault EJ, Kirsch RF, Acosta AM (1999) Multiple-input, multiple-output system identification for characterization of limb stiffness dynamics. Biol Cybern 80(5):327–337
29. Perreault EJ, Kirsch RF, Crago PE (2001) Effects of voluntary force generation on the elastic components of endpoint stiffness. Exp Brain Res 141(3):312–323
30. Perreault EJ, Kirsch RF, Crago PE (2002) Voluntary control of static endpoint stiffness during force regulation tasks. J Neurophysiol 87(6):2808–2816
31. Pierre M, Kirsch R (2002) Measuring dynamic characteristics of the human arm in three dimensional space. In: Proceedings of the 24th annual conference and the annual fall meeting of the Biomedical Engineering Society EMBS/BMES conference, engineering in medicine and biology, vol 3, pp 2558–2560
32. Pongrac H, Hinterseer P, Kammerl J, Steinbach E, Färber B (2006) Limitations of human 3D-force discrimination. In: Proceedings of the second international workshop on human-centered robotics systems
33. Pruszynski JA, Kurtzer I, Scott SH (2011) The long-latency reflex is composed of at least two functionally independent processes. J Neurophysiol 106(1):449–459
34. Rack PMH, Westbury DR (1974) The short range stiffness of active Mammalian muscle and its effect on mechanical properties. J Physiol 240(2):331–350
35. Sabes PN, Jordan MI, Wolpert DM (1998) The role of inertial sensitivity in motor planning. J Neurosci 18(15):5948–5957
36. Schouten AC, de Vlugt E, van Hilten JJB, van der Helm FCT (2008) Quantifying proprioceptive reflexes during position control of the human arm. IEEE Trans Biomed Eng 55(1):311–321

37. Shadmehr R, Mussa-Ivaldi F, Bizzi E (1993) Postural force fields of the human arm and their role in generating multijoint movements. J Neurosci 13(1):45–62
38. Shergill S, Bays P, Frith C, Wolpert D (2003) Two eyes for an eye: the neuroscience of force escalation. Science 301(5630):187
39. Tanaka Y, Tsuji T (2008) Directional properties of human hand force perception in the maintenance of arm posture. In: Ishikawa M, Doya K, Miyamoto H, Yamakawa T (eds) Neural information processing. Lecture notes in computer science, vol 4984. Springer, Berlin/Heidelberg, pp 933–942
40. Trumbower RD, Krutky MA, Yang BS, Perreault EJ (2009) Use of self-selected postures to regulate multi-joint stiffness during unconstrained tasks. PLoS ONE 4(5):e5411
41. Tsuji T, Morasso P, Goto K, Ito K (1995) Human hand impedance characteristics during maintained posture. Biol Cybern 72(6):475–485
42. Van Beek FE, Bergmann Tiest WM, Kappers AML (2013) Anisotropy in the haptic perception of force direction and magnitude. IEEE Trans Haptics 6(4):399–407
43. Van der Helm FCT, Schouten AC, de Vlugt E, Brouwn GG (2002) Identification of intrinsic and reflexive components of human arm dynamics during postural control. J Neurosci Methods 119(1):1–14
44. Van der Linde R, Lammertse P (2003) HapticMaster – a generic force controlled robot for human interaction. Ind Robot Int J 30(6):515–524
45. Vicentini M, Galvan S, Botturi D, Fiorini P (2010) Evaluation of force and torque magnitude discrimination thresholds on the human hand-arm system. ACM Trans Appl Percept 8(1):1–16
46. Yang XD, Bischof W, Boulanger P (2008) Perception of haptic force magnitude during hand movements. In: Proceedings of the IEEE international conference on robotics and automation, pp 2061–2066

Part II
Dynamic Perception

Chapter 5
Discrimination of Distance

Abstract While quite some research has focussed on the accuracy of haptic perception of distance, information on the precision of haptic perception of distance is still scarce, particularly regarding distances perceived by making arm movements. In this study, eight conditions were measured to answer four main questions, which are: what is the influence of reference distance, movement axis, perceptual mode (active or passive) and stimulus type on the precision of this kind of distance perception? A discrimination experiment was performed with 12 participants. The participants were presented with two distances, using either a haptic device or a real stimulus. Participants compared the distances by moving their hand from a start to an end position. They were then asked to judge which of the distances was the longer, from which the discrimination threshold was determined for each participant and condition. The precision was influenced by reference distance. No effect of movement axis was found. The precision was higher for active than for passive movements and it was a bit lower for real stimuli than for rendered stimuli, but it was not affected by adding cutaneous information. Overall, the Weber fraction for the active perception of a distance of 25 or 35 cm was about 11% for all cardinal axes. The recorded position data suggest that participants, in order to be able to judge which distance was the longer, tried to produce similar speed profiles in both movements. This knowledge could be useful in the design of haptic devices.

Previously published as:
F.E. van Beek, W.M. Bergmann Tiest & A.M.L Kappers (2014)
Haptic discrimination of distance
PLoS ONE, 9(8):e104769.

5.1 Introduction

Humans often have to perceive to which location they move their arm, for instance when reaching for the light switch in the dark. Usually, these kinds of tasks are a combination of the perception of distance and position [3, 19, 25, 32]. In general,

F.E. van Beek, *Making Sense of Haptics*, Springer Series on Touch and Haptic Systems, https://doi.org/10.1007/978-3-319-69920-2_5

reproducing a position yields a movement ending closer to the physical location than reproducing a distance [11, 16, 17]. The former, perception of position, has received a lot of attention in, for instance, work on the motor system (e.g. [30, 43]). In this article, we focus on the latter, the haptic perception of distance. Distance can be perceived in two different ways. Firstly, it can be perceived by exploring the length of a hand-held object. For this type of exploration, the finger span-method is often used, which involves the perception of the length of an object that is pinched between the thumb and index finger [1, 9, 22, 39]. Secondly, distance can be perceived by moving the hand over a certain distance, which involves tracing an object or moving over a well-defined path between a start and end position [4, 11, 17, 48]. Like all forms of measurements, perception can be described using the terms perceptual *accuracy* (also called constant error or bias) and *precision* (also called random error or discrimination threshold). Most studies on haptic distance perception have been focussed on perceptual accuracy, while precision has received hardly any attention, especially in the case of the movement method. In this article, we would like to extend the knowledge on the precision of distance perception using the movement method. Precise arm movements can be very important in, for instance, applications like haptic devices. Moreover, from a fundamental point of view, knowledge on perceptual precision provides a measure of the repeatability of the data obtained in these kinds of experiments. This can be a valuable addition to data on perceptual accuracy, which describe biases in perception.

We try to answer four main questions, which are: what are the effects of reference distance, movement axis, movement mode (active or passive), and stimulus type on the precision of haptic perception of distance? In the following paragraphs the existing knowledge on the effects of the four conditions on both aspects (accuracy and precision) of haptic perception of distance will be described.

5.1.1 Reference Distance

For the finger-span method, which can be used to perceive distances of hand-held objects, the relation between physical length and perceived length is a power function with an exponent ranging between 1.1 and 1.3 [21, 37, 41, 45]. For the movement method, a power function with an exponent of about 0.89–1 is reported [23, 40]. In general, short distances are underestimated and long distances are overestimated (e.g. [35]).

The precision of length perception as a function of reference length has been studied for the finger-span method. Gaydos [13] reports a stable Weber fraction (Wf: the smallest perceivable difference – also called discrimination threshold – divided by the stimulus intensity) of about 4% for reference lengths larger than 35 mm. Below that length, the Weber fraction increases up to 10% for a reference length of 10 mm [9, 37, 39]. However, nothing is known about the influence of reference distance on the precision of haptic distance perception using the movement method.

5.1.2 Movement Axis

For the accuracy of haptic distance perception using arm movements, an anisotropy between movements along different axes exists. This anisotropy is called the radial-tangential or the horizontal-vertical illusion (for a review, see Gentaz and Hatwell [14, 15]) and it entails that the radial (vertical) segment of the figure is perceived as longer than the tangential (horizontal) segment [2, 6, 48]. There have been numerous studies and debates concerning the influence of particular task characteristics on the presence and size of the effect. Generally, it is found that a distance is overestimated (underestimated) when the arm is moved at a slower (faster) speed [18, 24, 46] and as radial movements are indeed executed slower, this might be an explanation for the anisotropy [2, 48]. Recently, however, MacFarland and Soechting [26] systematically manipulated arm speed and effort of participants judging radial and tangential distances and found no effect of either manipulation on the size of the illusion. Other authors have shown that the illusion is still present when the L-shape is presented at an angle with the radial and tangential axes [8, 12, 33], but it disappears when the stimulus is presented in the vertical (fronto-parallel) plane [7, 8].

The question remains whether this anisotropy is also present in the precision of distance perception. Apart from early work, presenting data of only one subject for the comparison of two distances [20, 24], no data on precision along different movement axes is available.

5.1.3 Movement Mode

Distance can be perceived either passively, by being guided over a distance or by moving a surface under a stationary finger [4, 26], or actively, by exploring the length of a hand-held object [1, 22] or by moving over a certain distance [11, 17, 48]. It seems that active perception provides a more accurate percept than passive perception [32], with the movement distance being slightly underestimated in the passive case [34, 36]. For a review on the difference between active and passive perception, see Symmons et al. [38].

Again, it is not known what the effect on precision is. This comparison could provide insight in the question whether distance perception is purely based on the start and end position of the movement, or also on the way in which the movement itself is made.

5.1.4 Stimulus Type

Some work on perceptual accuracy for different types of haptic stimuli has been performed. Noll and Weber [27] found that distances are considerably underestimated when perceived purely cutaneously by moving a medium under the finger. When the finger is moved over the medium, thus providing cutaneous and kinaesthetic cues, the underestimation is much smaller. Terada et al. [42] added a condition with only kinaesthetic cues and found the underestimation to be in-between that of the former two conditions for distances of 100 and 150 mm. Conversely, Van Doorn [44] report that for a stimulus length of 40 mm subjects were more accurate at judging line length when using cutaneous cues alone, compared to using kinaesthetic cues or a combination of the two.

Recently, Bergmann Tiest et al. [4] performed discrimination experiments that involved passive perception of distance by moving the hand over a distance, moving a surface under the static finger, and moving the finger over a static surface, all over a distance of 80 mm. They found Weber fractions of 25% for the stimulus moving under the finger and 11% for the other two conditions. It therefore seems that distance perception is possible when purely cutaneous information is present, but it improves when kinaesthetic information is added. However, the combination does not seem to be better than kinaesthetic information alone. It is still unknown what the contribution of the different cues in an active situation would be.

There are not many studies in which the distance perception of both real and rendered stimuli have been tested. One study reports that the precision of perceiving distances using a stylus to probe a surface is slightly better in a real than in a simulated environment [29]. Whether there is still a difference in perception when a stylus is not used, is unknown.

From the studies mentioned above it is apparent that the precision of haptic distance perception is still a largely unexplored area. In our study, we investigated the effects of movement axis, stimulus type, movement mode and reference distance on this precision.

5.2 Materials and Methods

5.2.1 Participants

Twelve naive participants took part in this study, 5 male and 7 female, aged 22 ± 3 years (mean $\pm$ standard deviation), with no known neurological disorders. Handedness was assessed using the Coren-test for handedness [5], which confirmed that all participants were right-handed. All participants gave written informed consent to participate in the study and received a small compensation for their time. Prior to the experiment, they were given written instructions on how to perform the experiment. This experiment was approved by the Ethics Committee of the Faculty of Human Movement Sciences (ECB).

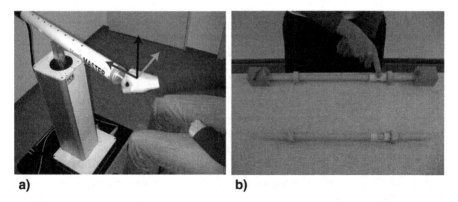

Fig. 5.1 Pictures of the setups used in the experiment. (**a**) The haptic device used for the conditions involving rendered stimuli, which were conditions 1 through 6. The three movement axes are indicated with arrows, light gray is the tangential, gray the radial and black the vertical axis. (**b**) The setup used for the conditions involving real stimuli, which were conditions 7 and 8. During the experiment, markers were taped on top of the blocks and on the nail of the participant's right index finger. These markers are not shown in this picture. The participant demonstrates condition 7, in which the finger is placed on the closed side of the carriage. The carriage at the other tube is placed with the open side up, as used in condition 8

5.2.2 Conditions

Eight conditions were measured in this experiment. The baseline condition was active distance perception, using the handle of a haptic device. In this condition, the reference distance was 25 cm, the movement axis was tangential to the participant in the horizontal plane and the haptic cues were of a kinaesthetic nature. The other conditions were variations in reference distance, movement axis, perceptual mode and stimulus type. The reference distance was either 15, 25 or 35 cm. The movement axis was either tangential, radial or vertical to the participant (see Fig. 5.1a). The perceptual mode was either active or passive. The type of stimulus was either rendered kinaesthetic, real kinaesthetic or real kinaesthetic+cutaneous. For an overview of all conditions, see Table 5.1.

5.2.3 Setup

The setup used for the rendered stimuli (conditions 1 through 6 in Table 5.1) was a 3 degrees of freedom haptic device, the Haptic Master (Moog Inc.). This is an admittance-controlled device, so the device is capable of simulating very stiff virtual objects. A virtual tunnel that was 2×2 mm wide was simulated, within which the participants could move a probe freely. The probe was a virtual point, which was located in the center of a ball-shaped handle (42 mm diameter). The handle was connected rigidly to the haptic device through a metal bar. Subjects were instructed

Table 5.1 Experimental conditions

Condition	Reference distance	Movement axis	Mode	Stimulus type
1	25 cm	Tangential	Active	Rendered
2	15 cm	Tangential	Active	Rendered
3	35 cm	Tangential	Active	Rendered
4	25 cm	Radial	Active	Rendered
5	25 cm	Vertical	Active	Rendered
6	25 cm	Tangential	Passive	Rendered
7	25 cm	Tangential	Active	Real
8	25 cm	Tangential	Active	Real+cutaneous

Overview of all the experiment conditions. Each of the four rightmost columns represents one research question

to grab the handle always in such a way that the metal bar was positioned between their index and middle finger and their palm was resting on top op the ball-shaped handle. For a top view of a participant holding the handle of the haptic device, see Fig. 5.1a. Participants could not move the probe out of the tunnel, as simulated walls with a stiffness of 20 kN/m prevented this. The length of the tunnel determined the movement distance. The position of the tunnel in space determined the start and end position. The orientation of the tunnel determined the movement axis. In the active condition, participants were asked to move their arm from the start to the end of the tunnel themselves. In the passive condition, the haptic device moved the arm of the participants.

The movement distances for conditions 1 through 6 were: a reference distance of 25 cm, and test distances of 21, 22, 23, 24, 26, 27, 28 and 29 cm. For the reference distances of 15 and 35 cm, the test distances ranged between 11 and 19 cm and between 31 and 39 cm, respectively, in steps of 1 cm. The method of constant stimuli was used, in which all test distances were presented 10 times for every condition, resulting in a total of 80 trials per condition. In all conditions, the start positions were offset either −2, 0 or +2 cm. The offset was assigned in a pseudo-random manner. The start position always differed between the first and the second movement of one trial.

In the tangential condition, the mean position halfway between the two stops was in front of the participant's sternum, while he or she made a movement from left to right. Participants were seated on a 62 cm high chair, while their arm height was 95 cm. In the radial condition, the height of the arm was the same, but the movement was along the radial axis in front of the sternum. In the vertical condition, the movement started at approximately the height of the other two conditions and was made upwards, again in front of the sternum. For an overview of the setup for the rendered conditions and the movement axes, see Fig. 5.1a.

For the conditions that involved real stimuli instead of rendered ones (conditions 7 and 8 in Table 5.1), another setup was used. This setup consisted of 27 PVC tubes with two stops each in between which a carriage could be moved (see Fig. 5.1b).

The carriage had two sides, one open and one closed, to create two different types of stimuli. The closed side of the carriage, which was used in condition 7, prevented the participants from feeling the surface sliding underneath their finger during the trial. In condition 8, the participant inserted his or her finger in a hole in the open side of the carriage to slide it over the surface of the tube while moving from start to end position. In both real conditions the arm movements, seat height, and arm height were the same as in the baseline condition. In conditions 1 through 7, the only types of relevant haptic cues were kinaesthetic. Therefore, the only difference between the rendered baseline condition and condition 7 was the hand posture (moving a handle vs. using the index finger to slide a carriage) and the type of stimulus (rendered vs. real). In condition 8, conversely, participants could also use cutaneous information.

During the trials of the conditions using rendered stimuli, hand position was measured using the position measurement function of the Haptic Master, which tracked the position of the probe, located at the center of the ball-shaped handle. In the conditions using real stimuli, the position of the finger was measured with a TrakSTAR device (Ascension Technology Corporation), which tracked the position of a marker on the nail of the right index finger using a magnetic field transmitter. Two extra markers were placed at the end blocks of the setup, to facilitate the data analysis. Both devices sampled position with a frequency of 90 Hz.

5.2.4 Procedure

Participants were blindfolded during the experiment. They participated in four 1-h sessions. During every session, two conditions were measured. At the start of every condition, three practice trials were performed. Every trial consisted of the comparison of two distances by moving once from start to end stop for every distance. During the movement, white noise was played on headphones worn by the participants to mask the sound of the device. A two-alternative forced-choice paradigm was used, so participants were only allowed to answer with '1' or '2' to indicate which distance they perceived to be the longer. Participants were provided with feedback on their answer, to direct them towards judging distance rather than position. Because the start position always differed between the two distances within one trial, participants could not use position cues directly and were forced to estimate distance. The order of start position and test distances was chosen pseudo-randomly and conditions were blocked. The order of the condition blocks was also chosen pseudo-randomly. Between conditions, participants were allowed to take a short break. Depending on the condition (see Table 5.1 for an overview), participants were asked to perform the task in a specified manner, as described below:

- Conditions 1 through 3: Participants moved the handle of the haptic device from left to right. At the right stop, they released the handle, after which it moved to the new start position. A sound and the start of the white noise indicated that

participants could start a new movement. They again moved the handle from left
to right and then indicated verbally which of the two distances they perceived to
be the longer, after which a new trial began.

- Condition 4: Participants moved the handle from proximal to distal. The rest of
  the procedure was the same as for conditions 1 through 3.
- Condition 5: Participants moved the handle from the most downwards to the
  most upwards position. The rest of the procedure was the same as for conditions
  1 through 3.
- Condition 6: After grabbing the handle, participants did not move it to the right
  themselves, but were moved by the device to the end position. The two distances
  within one trial were travelled with the same speed during half of the trials and
  had the same duration during the other half of the trials, in a pseudo-randomly
  assigned order. Mean movement speed was 0.167 m/s, mean movement duration
  was 90 s. The rest of the procedure was the same as for conditions 1 through 3.
- Conditions 7 and 8: After switching on the white noise, the experimenter placed
  the finger of the participant on the surface of the carriage (condition 7) or in the
  hole in the carriage (condition 8). The participant then moved the carriage until
  it hit the end stop. The experimenter replaced the rail with a rail with another
  distance, placed the carriage at the start position and switched the white noise on
  again, to indicate the start of the second part of the trial. The rest of the procedure
  was the same as for conditions 1 through 3.

5.2.5   Data Analysis

For each participant and condition, a psychometric curve was fitted to the answers
of the participants, using a least-squares fitting procedure on the following equation:

$$f(l) = \tfrac{1}{2} + \tfrac{1}{2}\mathrm{erf}\left(\tfrac{l - l_{\mathrm{ref}}}{\sqrt{2}\sigma}\right). \tag{5.1}$$

For a typical example of such a fit to the data, see Fig. 5.2. The σ in this equation
corresponds to the difference between the 0.50 and the 0.84 point, which is the
discrimination threshold that was used for further analysis. To calculate the Weber
fraction, the discrimination threshold was divided by the reference distance.

To investigate the effect of the various conditions on discrimination, a repeated
measures ANOVA was performed per research question. This resulted in four
ANOVAs for distance, movement axis, mode, and stimulus type, with condition as
the within-subject factor. When the sphericity criterion was violated, Greenhouse-
Geisser correction was used. Because the data set of the baseline condition was
used in each of the four ANOVAs, the α was Bonferroni-corrected by dividing it
by 4, so a p-value smaller than 0.0125 was deemed significant in this procedure.
When there was a significant main effect, post-hoc comparisons were performed

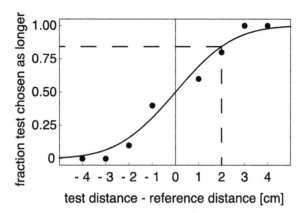

Fig. 5.2 Example of a psychometric curve fitted to the data of a single participant in a single condition. Black points represent mean values of ten trials with the same test distance. The curve is fitted using the error function described in Eq. 5.1. Note that this function forces the fraction to be 0.5 when the test distance equals the reference distance. The discrimination threshold is the fitted σ in the function, which is the value on the horizontal axis that corresponds to a fraction of 0.84. For the data in this figure, the fitted σ is 1.98 cm, which is indicated by the dashed lines

using Bonferroni-correction. The corrected α was based on the total number of post-hoc tests that were performed, which was 4, so a p-value smaller than 0.0125 was deemed significant for the post-hoc tests.

The data of the passive condition were further analyzed by dividing each set into a set with trials with the same movement speed and a set with the same movement time. To these two data sets per participant, new psychometric curves were fitted. The acquired thresholds were compared using a paired t-test.

From the position data, velocities were calculated for all active conditions. These were low-pass filtered using an 11-sample moving average, which yielded speed profiles like the example shown in Fig. 5.3. From these velocity profiles, the peak speed and end speed and their moment in time were calculated. This yielded the following parameters: peak speed, time of peak speed, end speed, and time of end speed. Within each trial, the difference between the parameters was calculated by subtracting the parameters for the first movement from that of the second. These difference parameters were then divided into two groups, based on the answer that the participant had given to the question which distance was the longer. For each condition and participant, the values were averaged over all trials. From these means per condition, a mean per participant per group was calculated for each parameter. The difference between the groups was assessed per parameter using a paired t-test. For one of the participants there was a measurement error in the position data of condition 7. The data of this condition were therefore not analyzed for this participant. One other participant was found to consistently have started his movement a little before the sound had indicated that he could start his trial. Because the data recording started simultaneously with the sound, the position for the first

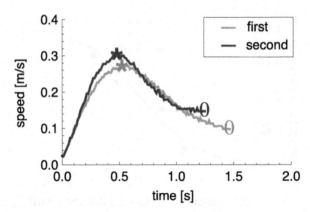

Fig. 5.3 Typical example of speed data of the first (light gray) and the second (dark gray) movement of one trial. The asterisks show the moment of peak speed, while ovals indicate the end of the movement. For each trial, the horizontal (for time) and vertical (for speed) distances from the axes to the asterisks and ovals were calculated to determine the speed parameters. The speed difference parameters were then calculated by subtracting each parameter of the first trial from that of the second trial. Data collection was stopped immediately when the participant reached the end position, therefore the end speed is not zero

part of each trial was not recorded for this participant. Therefore, the position data of this participant were not used in the analysis. Overall, the position parameters were thus based on the mean value of 6 conditions for one participant and on the mean values of 7 conditions for the remaining 10 participants.

5.3   Results

An overview of the discrimination thresholds for all conditions can be found in Fig. 5.4. Each group of bars in the figure answers one of the research questions. The 'distance' group showed mean thresholds of 2.1 ± 0.1, 2.8 ± 0.3 and 3.8 ± 0.4 cm (mean $\pm$ standard error of mean) for a reference distance of 15, 25 and 35 cm, respectively. The ANOVA showed a significant main effect of condition ($F_{2,22} = 18, p < 0.001$). Post-hoc comparisons showed that all conditions within this group differed significantly (15–25 cm: $p = 0.009$, 25–35 cm: $p = 0.009$, 15–35 cm: $p < 0.001$). When expressed as Weber fractions by dividing the thresholds by the reference distances, the fractions for the 'distance' group were 14.3 ± 0.7, 11.2 ± 1.0, and 10.8 ± 1.0 % for a reference distance of 15, 25 and 35 cm, respectively (see Fig. 5.5). In this case, there was a main effect of reference distance ($F_{2,22} = 9.0, p = 0.001$), but only the Weber fraction for a reference distance of 15 cm was significantly different from those for 25 and 35 cm ($p = 0.011$ and $p = 0.005$, respectively). The 'axis' group showed mean thresholds for tangential, radial and vertical movements of 2.8 ± 0.3, 2.9 ± 0.2 and 2.9 ± 0.2 cm, respectively.

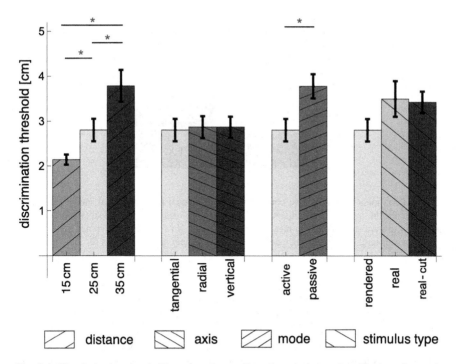

Fig. 5.4 Discrimination thresholds, grouped according to research question. The bars denote the mean of all participants and the error bars show the standard error of the mean of the participants. The labels at the horizontal axis indicate the measured conditions. Note that in each group the gray bar without hatching represents the baseline condition. See Table 5.1 for an explanation of all the conditions. $* = p < 0.0125$ (Bonferroni-corrected α)

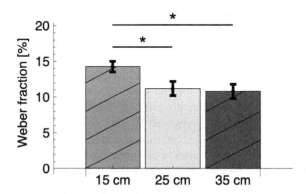

Fig. 5.5 Precision for the three reference distances, represented as Weber fractions. The Weber fraction is calculated by dividing the discrimination threshold by the reference distance. The bars show the mean over participants and the error bars show the standard error of the mean over participants. The Weber fraction for 15 cm is slightly larger than the fractions for 25 and 35 cm. $* = p < 0.0167$ (Bonferroni-corrected α)

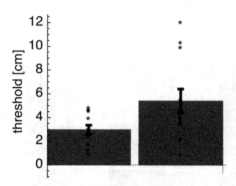

Fig. 5.6 Thresholds for the passive condition, split into equal time and equal speed trials. The asterisks mark the mean per participant, the bars show the mean over participants and the error bars show the standard error of the mean over participants. No significant difference between the 2 groups was found, but some participants did show extremely high thresholds for the equal time group

These thresholds were not significantly different ($F_{2,22} = 0.03, p = 0.971$). The 'mode' group showed mean thresholds of 2.8 ± 0.3 and 3.8 ± 0.3 cm for the active and the passive condition, respectively. These thresholds were significantly different ($F_{1,11} = 15, p = 0.003$). The 'stimulus type' group showed mean thresholds of 2.8 ± 0.3, 3.5 ± 0.4 and 3.4 ± 0.2 cm for the rendered, real and real+cutaneous condition, respectively. These thresholds were not significantly different ($F_{1.1,13} = 3.1, p = 0.099$).

For the grouping into equal speed and time trials for the passive condition, thresholds seem a bit lower when the speed of the first and second movement was the same, compared to trials in which the movement time was the same (see Fig. 5.6). However, they were not significantly different ($t_{11} = -2.1, p = 0.056$).

The velocity profiles that were constructed from the position data (for a typical example, see Fig. 5.3) showed quite some differences between the cases in which participants had answered '1' and cases in which they had answered '2' to be the longer distance. These differences are shown in Fig. 5.7. For the time parameters, there was a significant difference for end time ($t_{10} = -8.0, p < 0.001$), but the time of peak speed was not significantly different. For end time, the time difference was negative when participants answered '1' and positive when they answered '2', meaning that the time of the first trial was longer than that of the second when they answered '1' and shorter when they answered '2'. For both speed parameters, there were significant differences between the two answers, which were $t_{10} = 8.6, p < 0.001$ for peak speed and $t_{10} = 5.5, p < 0.001$ for end speed. Both speed differences were positive when participants answered '1' and negative when they answered '2', meaning that the speed was lower for the first stimulus of a trial than for the second stimulus when participants answered '1', while it was higher when they answered '2'.

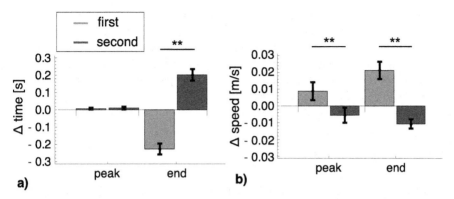

Fig. 5.7 Position data, grouped according to answers of participants, with light gray for '1' and dark gray for '2'. All parameters were calculated by subtracting the parameter of the first trial from that of the second trial. The bars show the mean over participants and the error bars show the standard error of the mean over participants. (**a**) Time parameters, which are: time to peak speed and total movement time. (**b**) Speed parameters, which are: peak speed and end speed. ** $= p < 0.01$

5.4 Discussion

From the discrimination experiments, an influence of reference distance and perception mode on the precision of haptic distance perception was observed (see Fig. 5.4). Below, the results for each research question will be discussed in more detail. For each question, the possible implications of our findings for the design of haptic devices will also be discussed. However, these implications are only valid when optimizing for precision of distance perception.

5.4.1 Reference Distance

The discrimination threshold was influenced by reference distance. From Weber's law [47], a constant Weber fraction (and thus an increasing absolute threshold with increasing reference distance) could be expected. However, the Weber fraction was slightly higher for a reference distance of 15 cm than for the other two distances (Fig. 5.5). This is in line with results for the finger-span method, which also show an increasing Weber fraction for very small distances [9, 37, 39]. Probably, a whole arm movement is not the most efficient way to estimate a distance of 15 cm. This distance is only a bit longer than the span of one hand, so moving over this distance generates only a small difference in joint angles. For the two largest movement distances, a Weber fraction of about 11% was found. Thresholds for the finger-span method for intermediate distances were at least twice as small, so it is easier to precisely perceive distances by perceiving distance between thumb and forefinger, than by whole arm movements. To allow users of haptic devices to grasp objects between thumb and forefinger, the interfaces should be designed in such a way that operators can use their individual fingers and can receive feedback on them. When this is possible, operators can use the finger-span method to perceive object size for small distances. This would increase their precision, compared to using the movement method with a handle held in a power grip.

5.4.2   Movement Axis

We found no effect of movement axis on the precision of distance perception, while there is a well-known effect of movement axis on the accuracy of distance perception, called the radial-tangential illusion [2]. Generally, the distance of a radial movement is overestimated, compared to a tangential one. However, a difference in accuracy does not automatically imply a difference in precision. Imagine, for instance, that a participant perceives the same physical distance (e.g. 10 cm) as twice as large along the radial axis (20 cm), compared to the tangential axis (10 cm). This participant might perceptually also need a twice as large difference between two distances presented along the radial axis (e.g. 2 cm) to perceive them as being different, compared to two distances presented along the tangential axis (1 cm). However, the thresholds are measured in the physical world, so for this participant no difference in precision between the two directions will be found (both 1 cm). For more details on this topic, see the review by Ross [31]. Our measured discrimination thresholds did not differ between the cardinal axes, so we found no indication of an influence of the radial-tangential illusion on the precision of distance perception. As visual perception of depth (along the radial axis) is generally less precise than visual perception in the fronto-parallel plane (along the tangential and vertical axes [28]), haptic depth cues could potentially aid in precise perception of distances using devices that combine haptic and visual information.

5.4.3   Movement Mode

For the comparison between movement modes, we found a deterioration of precision for the passive condition, which is in line with the results on accuracy of distance perception [32]. This is an interesting observation, because it suggests that the perception of distance is not solely based on the position of the start and end points, but the way in which the movement in between the positions is made also seems to add information. The grouping of the data set into trials with the same speed and the same movement time (see Fig. 5.6) was made to assess whether participants are more likely to have been using speed differences or time differences in this task. Evidence from this grouping is not conclusive, as no significant differences were found, but it suggests that some participants were relying more on time cues than on speed cues, judged from extremely high thresholds for some participants in the equal time group. This finding suggests that taking away authority from the operator, for instance by providing strong guidance forces in haptic devices, could deteriorate precision of distance perception.

5.4.4   Stimulus Type

We found no effect of stimulus type. Although the thresholds are not significantly different, they do look a bit higher for the real conditions than for the rendered one. Usually, differences between stimulus types are in favor of real stimuli (e.g. [29]). Both the difference in hand posture (sliding a finger over a tube in the conditions using real stimuli and holding a handle in the conditions using rendered ones) and the difference between the stimulus properties (real and rendered) could be the cause of this. We suspect that the stimulus properties might be important, because the movement with the real stimulus was a bit less constrained than the movement with the rendered stimulus. Because the tubes were round, the carriage also had a little freedom to rotate around the tube, which could have resulted in a forward-backward movement. In the conditions using rendered stimuli, the maximum movement amplitude in this direction was 2 mm, which was the width of the haptic tunnel. For the conditions using real stimuli, position data were used to calculate the maximum movement amplitude in this direction per trial, which yielded a mean amplitude maximum of 5.8 mm over all trials. This could explain why the thresholds for the real condition look a bit higher. However, it is intriguing that the addition of cutaneous information, which was done in condition 8, did not seem to help the participants. From an optimal cue combination perspective [10], at least a little bit of improvement should be expected. It seems therefore that in our task, cutaneous information was so unreliable that it did not add much as a predictor in the cue combination model. Bergmann Tiest et al. [4] show that purely cutaneous distance discrimination is possible, but is a lot less precise than kinaesthetic perception. In their experiment, which was a passive perception experiment, there was also hardly any added value of combining cutaneous and kinaesthetic information. In our active case, the same principle seems to hold. This information seems to imply that for haptic devices it is not necessary to render surfaces that resemble real surfaces to ensure precise distance perception.

5.4.5   Movement Strategy

The grouping of the passive data into equal speed and equal time trials did not yield significant differences. However, the trend seems to suggest that some participants were more likely to use an estimate of the movement time than an estimate of movement speed as a strategy to find out which distance was the longer. For the active conditions, participants could use their own movement strategy to obtain the best estimate of movement distance. To get an insight into the strategy that the participants were using in this active case, the position data were analyzed (Fig. 5.7).

For both speed and time data significant differences between answering '1' and '2' were found. This can be understood by looking at the data qualitatively (for an example of the shape of the curves, see Fig. 5.3), as the velocity profiles of the movements within one trial look very much alike. It therefore seems that participants tried to reproduce the speed profile of the first trial during the second trial. This would be a smart strategy, as simply judging whether the end point is reached before or after the reproduced profile is finished, would give the participants all the information they need to answer the question. If we assume that this was indeed the strategy that participants were using, we can speculate on its effect on the parameters. For the time data (shown in Fig. 5.7a), the parameters are congruent with our explanation, as the peak times did not differ, while the end times did. When participants answered '1', the sign of the end time difference was negative, which means that the first trial took longer when they judged it to be the longer distance. For the speed data (shown in Fig. 5.7b), both parameters differed significantly between the answers. For end speed, the sign makes sense: when participants answered '1', the sign of the end speed difference was positive, meaning that the speed in the second trial was still higher at the end. This indicates that participants could not complete their profile in the second trial and thus judged this distance to be the shorter one. For peak speed, however, a significant difference was also found, so the reproduction of the speed profiles was not perfect. When, for instance, the peak speed was higher during the second trial, this would result in a smaller end time and a higher end speed in the second trial, which would induce participants to judge the first distance to be the longer. This is indeed what the parameters reflect: when the peak speed in the second trial was higher, the participants answered '1' more often. So, this imperfect reproduction of the peak speed seems to have influenced the perception of the participants.

Concluding, the parameters can be explained by assuming that the participants tried to reproduce the speed profile of the first trial during the second trial and based their decision on whether they could complete their profile or not. Apparently, they succeeded in reaching the peak speed at the same moment, but they did not succeed in reaching exactly the same peak speed magnitude. Of course, this reasoning is based on speculation, but it does explain all the measured parameters.

5.4.6   Conclusion

Overall, we found that for movements along distances of 25 and 35 cm, a Weber fraction of about 11% was reached for the precision of active haptic distance perception along all cardinal axes. Passive movements worsen the precision, while adding cutaneous information does not improve it.

References

1. Abravanel E (1971) The synthesis of length within and between perceptual systems. Percept Psychophys 9(4):327–328
2. Armstrong L, Marks L (1999) Haptic perception of linear extent. Percept Psychophys 61(6):1211–1226
3. Ashby A, Shea C, Howard RM (1980) Short term memory for kinesthetic movement information: influence of location cues on recall of distance. Percept Mot Skills 51:403–406
4. Bergmann Tiest WM, van der Hoff LMA, Kappers AML (2011) Cutaneous and kinaesthetic perception of traversed distance. In: Proceedings of the IEEE world haptics conference, pp 593–597
5. Coren S (1993) The left-hander syndrome. Vintage Books, New York
6. Davidon RS, Cheng MFH (1964) Apparent distance in a horizontal plane with tactile-kinesthetic stimuli. Q J Exp Psychol 16(3):277–281
7. Day RH, Avery GC (1970) Absence of the horizontal-vertical illusion in haptic space. J Exp Psychol 83(1):172–173
8. Deregowski J, Ellis H (1972) Effect of stimulus orientation upon haptic perception of the horizontal-vertical illusion. J Exp Psychol 95(1):14–19
9. Durlach N, Delhorne L, Wong A, Ko W, Rabinowitz W, Hollerbach J (1989) Manual discrimination and identification of length by the finger-span method. Percept Psychophys 46(1):29–38
10. Ernst MO, Banks MS (2002) Humans integrate visual and haptic information in a statistically optimal fashion. Nature 415(6870):429–433
11. Faineteau H, Gentaz E, Viviani P (2003) The kinaesthetic perception of euclidean distance: a study of the detour effect. Exp Brain Res 152(2):166–172
12. Fasse ED, Hogan N, Kay BA, Mussa-Ivaldi FA (2000) Haptic interaction with virtual objects. Biol Cybern 82(1):69–83
13. Gaydos HF (1958) Sensitivity in the judgment of size by finger-span. Am J Psychol 71(3):557–562
14. Gentaz E, Hatwell Y (2004) Geometrical haptic illusions: the role of exploration in the Müller-Lyer, vertical-horizontal, and Delboeuf illusions. Psychon Bull Rev 11(1):31–40
15. Gentaz E, Hatwell Y (2008) Haptic perceptual illusions. In: Grunwald M (ed) Human haptic perception: basics and applications. Birkhäuser, Basel, pp 223–233
16. Gupta RK, Gupta M, Kool VK (1986) Role of vision and kinesthesis in location and distance estimates. Acta Psychologica 62(2):141–159
17. Hermelin B, O'Connor N (1975) Location and distance estimates by blind and sighted children. Q J Exp Psychol 27(2):295–301
18. Hollins M, Goble A (1988) Perception of length of voluntary movements. Somatosens Res 5(4):335–348
19. Imanaka K (1989) Effect of starting position on reproduction of movement: further evidence of interference between location and distance information. Percept Mot Skills 68(2):423–434
20. Jastrow J (1886) The perception of space by disparate senses. Mind 11(44):539–554
21. Jones B (1983) Psychological analyses of haptic and haptic-visual judgements of extent. Q J Exp Psychol Sect A 35(4):597–606
22. Kelvin RP, Mulik A (1958) Discrimination of length by sight and touch. Q J Exp Psychol 10(4):187–192
23. Lanca M, Bryant DJ (1995) Effect of orientation in haptic reproduction of line length. Percept Mot Skills 80(3c):1291–1298
24. Leuba JH (1909) The influence of the duration and of the rate of arm movements upon the judgment of their length. Am J Psychol 20(3):374–385
25. Marteniuk RG, Roy EA (1972) The codability of kinesthetic location and distance information. Acta Psychologica 36(6):471–479

26. McFarland J, Soechting JF (2007) Factors influencing the radial-tangential illusion in haptic perception. Exp Brain Res 178(2):216–227
27. Noll NC, Weber RJ (1985) Visual and tactile scanning: moving scan versus moving medium. Bull Psychon Soc 23(6):473–476
28. Norman JF, Todd JT, Perotti VJ, Tittle JS (1996) The visual perception of three-dimensional length. J Exp Psychol Hum Percept Perform 22(1):173–186
29. O'Malley MK, Goldfarb M (2005) On the ability of humans to haptically identify and discriminate real and simulated objects. Presence Teleop Virt 14(3):366–376
30. Rincon-Gonzalez L, Buneo CA, Helms Tillery SI (2011) The proprioceptive map of the arm is systematic and stable, but idiosyncratic. PLoS ONE 6(11):e25214
31. Ross HE (1997) On the possible relations between discriminability and apparent magnitude. Br J Math Stat Psychol 50(2):187–203
32. Roy EA, Diewert GL (1975) Encoding of kinesthetic extent information. Percept Psychophys 17(6):559–564
33. Soechting JF, Flanders M (2011) Multiple factors underlying haptic perception of length and orientation. IEEE Trans Haptic 4(4):263–272
34. Stelmach GE, Kelso JS (1975) Memory trace strength and response biasing in short-term motor memory. Mem Cogn 3(1):58–62
35. Stelmach GE, Wilson M (1970) Kinesthetic retention, movement extent, and information processing. J Exp Psychol 85(3):425–430
36. Stelmach GE, Kelso JS, McCullagh PD (1976) Preselection and response biasing in short-term motor memory. Mem Cogn 4(1):62–66
37. Stevens S, Stone G (1959) Finger span: ratio scale, category scale, and JND scale. J Exp Psychol 57(2):91–95
38. Symmons M, Richardson B, Wuillemin D (2004) Active versus passive touch: superiority depends more on the task than the mode. In: Ballesteros S, Heller M (eds) Touch, blindness, and neuroscience. UNED Press, Madrid, pp 179–185
39. Tan HZ, Pang XD, Durlach NI (1992) Manual resolution of length, force and compliance. In: Kazerooni H (ed) Advances in robotics, vol 42. The American Society of Mechanical Engineers, New York, pp 13–18
40. Teghtsoonian M, Teghtsoonian R (1965) Seen and felt length. Psychon Sci 3:465–466
41. Teghtsoonian R, Teghtsoonian M (1970) Two varieties of perceived length. Percept Psychophys 8(6):389–392
42. Terada K, Kumazaki A, Miyata D, Ito A (2006) Haptic length display based on cutaneous-proprioceptive integration. J Rob Mechatron 18(4):489–498
43. van Beers RJ, Sittig AC, Denier van der Gon JJ (1998) The precision of proprioceptive position sense. Exp Brain Res 122(4):367–377
44. van Doorn GH, Richardson BL, Symmons MA, Howell JL (2012) Cutaneous inputs yield judgments of line length that are equal to, or better than, those based on kinesthetic inputs. In: Isokoski P, Springare J (eds) Haptics: perception, devices, mobility, and communication. Lecture notes in computer science, vol 7283. Springer, Berlin/Heidelberg, pp 25–30
45. van Doren CL (1995) Cross-modality matches of finger span and line length. Percept Psychophys 57(4):555–568
46. Wapner S, Weinberg J, Glick J, Rand G (1967) Effect of speed of movement on tactual-kinesthetic perception of extent. Am J Psychol 80(4):608–613
47. Weber E (1978/1834) De tactu. In: E.H. Weber on the tactile senses. Erlbaum (UK) Taylor & Francis, Hove
48. Wong TS (1977) Dynamic properties of radial and tangential movements as determinants of the haptic horizontal-vertical illusion with an L figure. J Exp Psychol Hum Percept Perform 3(1):151–164

Chapter 6
The Effect of Damping on the Perception of Hardness

Abstract In teleoperation systems, damping is often injected to guarantee system stability during contact with hard objects. In this study, we used psychophysical experiments to assess the effect of adding damping on the user's perception of object hardness. In Experiments 1 and 2, combinations of stiffness and damping were tested to assess their effect on perceived hardness. In both experiments, two tasks were used: an in-contact task, starting at the object's surface, and a contact-transition task, including a free-air movement. In Experiment 3, the difference between global damping (present throughout the environment) and local damping (present inside the object only) was tested. In all experiments, force and position data were recorded to assess which parameters correlated with the participant's perceptual decision. Experiments 1 and 2 show that with added damping, perceived hardness increased for an in-contact task, while it decreased for a contact-transition task, with the latter effect being much larger than the former. Experiment 3 shows that this effect was mainly due to the addition of global damping, since there was a large perceptual difference between adding global and local damping. The force and position parameters show that object indentation, mean velocity and adjusted rate-hardness correlated most strongly with the participant's perceptual experience.

Previously published as:
F.E. van Beek, D.J.F. Heck, H. Nijmeijer, W.M. Bergmann Tiest & A.M.L. Kappers
The effect of global and local googledamping on the perception of hardness
IEEE Transactions on Haptics 9(3), 409–420
DOI: 10.1109/TOH.2016.2567395.

Parts of the data in this chapter have also been published in:
F.E. van Beek, D.J.F. Heck, H. Nijmeijer, W.M. Bergmann Tiest & A.M.L. Kappers (2015)
The effect of damping on the perception of hardness
Proceedings of the 2015 IEEE World Haptics Conference (WHC), 82–87
DOI: 10.1109/WHC.2015.7177695.

6.1   Introduction

New technologies like deep-sea mining, remote handling for nuclear technologies and space applications rely heavily on the use of teleoperation systems. In these systems, the operator uses a master device to control a slave system in the remote environment. By establishing a feedback loop from the remote environment to the operator, a bilateral connection is created and the operator can be provided with force feedback from the remote environment. If the quality of the feedback is very good, which typically implies that the impedance of the remote environment is reflected properly through the force feedback signal, the operator is provided with a very accurate representation of the remote environment [11]. An important factor determining the feedback quality is the choice of the control architecture governing the bilateral connection, which depends on several system properties.

One of those properties is the presence of delays in the system [17]. If the master and slave are physically separated over a large distance, there is a delay between sending information at one end and receiving it at the other end. These delays can cause instabilities in the system, which is an undesirable property, since they can damage the teleoperator or environment and prevent the operator from executing the task. Nuño et al. [13] have shown that one way to guarantee stability is to ensure that the ratio of injected damping by the controller and the proportional controller gain (which is strongly related to the stiffness reflected to the operator) is bounded from below. This bound, and thus the required amount of damping, increases for increasing delays. Hence, a common solution to prevent instabilities is to inject of a lot of damping or viscous friction into the system through the controller [see e.g. 8, 14, 19]. In this way, the stability is guaranteed in the presence of delays and, at least theoretically, it remains possible to reflect high stiffnesses, which would make the device potentially capable of providing high quality feedback. However, it is unclear what the effect of injecting damping is on the operator's perception of the remote environment. One of the important aspects conveyed by high quality haptic feedback is object hardness, since the proper assessment of hardness is crucial to identify objects and to properly interact with them. A surgeon, for instance, needs to feel the difference in hardness between the tumor that must be removed and the surrounding tissue that needs to stay intact. Moreover, the proper experience of hardness allows operators to exert the right amount of force to manipulate an object without damaging it. Therefore, the question in our study was: what is the effect of injecting damping on the perception of object hardness?

The perception of hardness of natural objects has been investigated quite thoroughly (see Bergmann Tiest [1] for a review), which has shown that stiffness plays a major role in making objects feel hard. The relationship between stiffness and perceived hardness of natural objects can be described by a power function with an exponent of $\sim$0.8 [7], while the Just Noticeable Difference (JND) for stiffness is $\sim$15% [3, 4]. For virtual objects, Lawrence et al. [12] have shown that *force rate of change at impact* divided by *displacement velocity at impact* seems to correlate better with perceived hardness than stiffness alone does. They called this new parameter 'rate-hardness'. Han and Choi [6] extended this work by describing a slightly different parameter, 'extended rate hardness', which they defined as *maximum force rate of change* divided by *displacement velocity at impact*, which should be suitable for a larger class of rendering algorithms. Lawrence et al. [12] stated that rate-hardness can be seen as a sum of stiffness and damping, leading to the final rate-hardness value. This underlines the potential importance of damping in the perception of hardness.

Rosenberg and Adelstein [15] have shown that directional dampers can provide a sensation of hardness and can feel 'wall-like', which suggests that damping can contribute to the perceived hardness of an object. However, the authors did not combine damping and stiffness in one object. Lawrence et al. [12] did use a control law to quickly adjust the object's stiffness, which should have resulted in something that resembles a combination of stiffness and damping. When limiting the interaction force to 2 N, they found that adding a damping-like parameter through this control law increased the perceived object hardness in a free exploration task. Han and Choi [6] essentially replicated the findings in a similar experiment using a comparable type of control law, while using a much stronger haptic device (maximum force 37.5 N) and a tapping task. However, in both studies the amount of damping was not controlled directly and therefore it is not known which amount of damping corresponds to which amount of increase in perceived hardness. Moreover, both studies only used damping inside the object, which we term *local damping*, while the injection of damping in control architectures for delayed bilateral teleoperation is typically done throughout the complete workspace, so both inside and outside of virtual objects, which we term *global damping*. So, the effects of global damping on the perception of hardness is still an open question.

In this study, we investigated the effect of damping on the perceived hardness of a virtual object, composed of a spring. The design mimics a common solution to guarantee stability in bilateral teleoperation, which is the injection of damping, and is therefore interesting for control engineers who aim to design high performance architectures. In Experiment 1, we studied the effect of *global* damping on the perception of hardness for various levels of stiffness. The results of Experiment 1 have been published previously in conference proceedings [21]. In Experiment 2, we studied the effect of *global* damping on the perception of hardness for various levels of damping. In both experiments, two types of tasks were used: (1) an in-contact task, in which participants started their movement at the object's surface and indented it, and (2) a contact-transition task, in which they first made a free-air movement, before making contact with nonzero impact velocity and indenting the object. In Experiment 3, the difference between the effect of *global* and *local* damping on the perception of hardness was investigated. In teleoperation systems, damping is usually implemented in a global fashion, so Experiments 1 and 2 were used to explore the stiffness-damping space that is relevant for these applications. Experiment 3 enabled us to compare the results from the first two experiments to those from literature on local damping. In the next section, the general methods are explained first, followed by experiment-specific paragraphs.

6.2   Material and Methods

6.2.1   Participants

In all experiments, 12 naive participants took part. None participated in more than 1 experiment. In Experiment 1, 3 males and 9 females took part. They were 21 ± 3 years old, 9 were right-handed and 3 were left-handed (assessed using the Coren-test for handedness [5]). In Experiment 2, 2 males and 10 females took part. They were 23 ± 3 years old, 11 were right-handed and 1 was left-handed. In Experiment 3, 4 males and 8 females took part. They were 28 ± 4 years old, 11 were right-handed and 1 was left-handed. None of the participants had a history of neurological disorders. Prior to the experiment, they received written instructions and signed an informed-consent form. For Experiments 1 and 2, they received a small payment for their participation. All the experiments were approved by the Ethics Committee of the Faculty of Human Movement Sciences (ECB).

6.2.2   Protocol

6.2.2.1   General

During all experiments, participants were sitting, while wearing a blindfold and headphones. The setup (see Fig. 6.1) was an admittance-controlled haptic device, the HapticMaster (Moog Inc.). This device is capable of rendering very high stiffnesses (maximum 20 kN/m) and large forces (maximum 250 N) [20]. The handle is a ball-shaped object, which is rigidly connected to the device through a metal bar. It had a simulated inertial mass of 3 kg, while its gravity was compensated for. Headphones worn by the subjects produced white noise to mask the sound of the haptic device.

Fig. 6.1 Blindfolded participant holding the handle of the HapticMaster. (a) Overview of the setup and force sensor. (b) Close-up of the hand posture of the participant

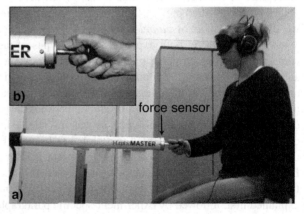

On each trial, participants were consecutively presented with two virtual objects, each composed of a linear spring. The participants were asked to grab the handle when the white noise started and then move the handle away from them along their sagittal axis, which was the axis to which the movement of the device was restricted. During their movement, they encountered the first object. They were free to move the handle into the virtual object as far as they liked, but they were only allowed to move forward. Once the participants felt they had reached the end of their forward movement, they had to release the handle and the white noise stopped. After this, the handle moved back to the start position, the white noise started again and the participant made the second movement to observe the second object. Once (s)he completed this movement, (s)he had to indicate which object felt harder. Subsequently, a new trial started. Both the start and the object were always located at the same positions. Figure 6.2 shows a top-view of the participant, the movement trajectory and the object.

To obtain a fast and precise measure of the difference between damped and undamped stimuli, a one-up-one-down staircase procedure was used in which the stiffness of the test stimulus was adapted. For an example of one of the staircases,

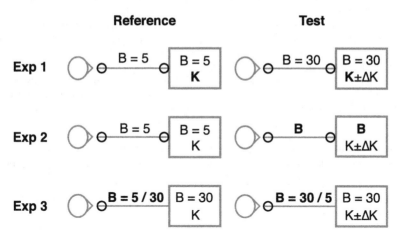

Fig. 6.2 Top view of the participant, the trajectory and the virtual object (all in gray) in the 3 experiments, for the reference and test stimulus. Participants always made movements away from their body, which is from left to right in the figure. In the in-contact task, the start position (indicated with the rightmost black circle for both stimuli in Experiments 1 and 2) was at the object's surface. For the contact-transition task, the participants started their movement 8 cm closer to their body (the leftmost black circle) and first made a free-air movement (indicated with the gray line) before encountering the object. The numbers show the stiffness (K) and damping (B) values of the reference and the test stimuli. Bold numbers indicate the property that was varied across conditions: in Experiment 1, various levels of stiffness were tested for 2 tasks; in Experiment 2, various levels of damping were tested for 2 tasks; in Experiment 3, local and global damping were compared for 1 task. In all experiments, the stiffness of the test stimulus was varied to find at which level the test and reference stimulus were perceptually equal in hardness

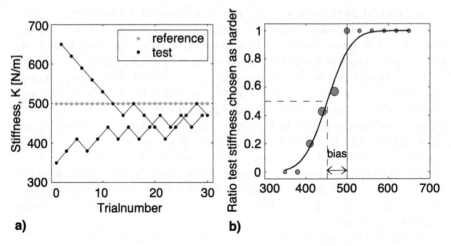

Fig. 6.3 Typical example of a staircase pair and the psychometric curve fitted to the perceptual data of one participant, for a reference stiffness of 500 N/m in an in-contact task in Experiment 1. (**a**) Staircase pair showing one experimental condition. In each trial, one reference stimulus (gray dot) and one test stimulus (black dot) was presented. If the participant perceived the test (reference) stimulus as the harder stimulus, the stiffness of the next test stimulus was decreased (increased) to find the test stiffness at which both stimuli were perceived as equally hard. (**b**) Psychometric function fitted to the perceptual data shown in panel (**a**). Gray dots show actual data points, while the black curve is the psychometric function fitted to the data. The size of the dots shows the number of times this point was measured and therefore its weight in the fit procedure. The vertical black line shows the reference stiffness. The PSE is the value on the horizontal axis corresponding to a ratio of 0.5, which is indicated with the dashed lines. The bias is the difference between the PSE and the reference stiffness, which is indicated with an arrow

see Fig. 6.3a. In a staircase procedure, the properties of the reference stimulus are kept constant, while the properties of the test stimulus are adjusted according to the answers of the participant. In our procedure, we used a reference stimulus with a constant level of stiffness and damping. The test stimulus had another constant damping level, while its test stiffness was adjusted during the experiment. In Experiments 1 and 2, we used 2 interleaved staircases in each condition, which we called a 'staircase pair'. In Experiment 3, we used two interleaved staircase pairs, so two conditions were measured simultaneously. In all experiments, each staircase pair consisted of one staircase with an initial stiffness that was 30% higher than that of the reference stimulus and one staircase with a initial stiffness that was 30% lower than the reference stiffness. In each trial, one reference and one test stimulus were presented. If the participant answered that the test (reference) stimulus was the harder stimulus, the test stiffness was decreased (increased) on the next trial to find the test stiffness at which the perceptual hardness of the test stimulus was equal to that of the reference stimulus. The step size with which the stiffness was changed was 6% of the reference stiffness for all experiments.

When both of the staircases of the staircase pair had reversed at least 5 times, both staircases were terminated. When the maximum of 50 trials was reached, both staircases were also terminated. In Experiments 2 and 3, we used a minimal number of 30 trials per staircase pair. In Experiment 1, this minimum was not set yet, but usually at least 30 trials were performed in Experiment 1 as well. The order of the conditions was counterbalanced between subjects, while the order of test and reference stimulus within each of the staircases was pseudo-randomly assigned, which ensured that both orders were tested equally often. Throughout the experiment, force and position data were recorded with a frequency of 1024 Hz using the datalogger function of the HapticMaster. Experiments 1 and 2 took about one hour per participant each, while Experiment 3 took about 30 min per participant.

6.2.2.2   Experiment 1

In Experiment 1, two types of tasks were used: a contact-transition task and an in-contact task. In the contact-transition task, participants first made a free-air movement of 8 cm towards the virtual object, before making contact with it. So, in this task, the movement consisted of two parts (while it was made as a single movement by the participant, such that the impact occurred with non-zero velocity): a free-air part and an in-contact part. It was stressed to the subjects, both verbally and in the written instruction, that they should not base their decision about the hardness of the virtual object on the free-air part, but rather focus on the impact and the part where they were in contact with the object. In the in-contact task, the free-air part of the movement was eliminated by placing the start position at the surface of the virtual object, so effectively only the in-contact part of the task remained. The rest of the instructions were the same for the two tasks. For a visualization of the task and the conditions, see Fig. 6.2.

For each task, 3 reference stiffnesses were used, which were 500, 1000 and 2000 N/m for the in-contact task and 1000, 2000 and 4000 N/m for the contact-transition task, resulting in 6 conditions (3 stiffness levels × 2 tasks). The stiffness ranges differed between the tasks, because we aimed to cover a broad range of hardnesses for both tasks. For the in-contact task, it was very difficult to indent an object with a stiffness above 2000 N/m. For the contact-transition task, it was very difficult to properly feel the transition from free-air movement to object indentation for a stiffness below 1000 N/m. Therefore, these different stiffness ranges were selected. On the trials where the reference stiffness was presented, the global damping was always 5 Ns/m. On trials where the test stiffness was presented, the global damping was always 30 Ns/m. We chose these values after preliminary testing, by selecting a damping value for the test stimulus which provided a considerable amount of damping, while still allowing the user to move comfortably. We did not design the reference stimulus to have no damping, because the device then sometimes became unstable upon object contact for high stiffnesses and velocities. We did include one condition in Experiment 2 in which the test stimulus was undamped, but this condition was not critical for the results.

6.2.2.3 Experiment 2

In Experiment 2, a contact-transition task and an in-contact task were used again. For the in-contact task, the reference stiffness was always 1000 N/m. For the contact-transition task, the reference stiffness was always 2000 N/m. In each condition, the global damping for the reference stimulus was always 5 Ns/m, while the global damping for the test stimulus differed between blocks: 0, 10, 15, 20, and 25 Ns/m were used, resulting in ten conditions (5 damping levels × 2 tasks). Figure 6.2 shows a visual description of these conditions.

6.2.2.4 Experiment 3

In Experiment 3, only a contact-transition task was used. The reference stiffness was always 2000 N/m. Two interleaved staircase pairs (4 staircases) were used to measure 2 conditions simultaneously. In both staircase pairs, one of the staircases consisted of locally damped stimuli (mainly damped inside the object), while the other consisted of globally damped stimuli (damped throughout the workspace). In both types of stimuli, the damping inside the object was always 30 Ns/m. For the globally damped stimuli, the situation was the same as for the test stimuli in Experiment 1: the damping outside of the object was also 30 Ns/m. For the locally damped stimuli, the damping outside of the object was 5 Ns/m. In one condition, the reference stimulus was damped locally, while the test stimulus was damped globally. In the other condition, the reference stimulus was damped globally, while the test stimulus was damped locally. So, the only difference between the conditions was which stimulus was the reference and the test. In Fig. 6.2 these conditions are illustrated visually. The experiment was measured in one block. The participants were asked if they wanted a break when the experiment was about halfway, but none of the participants felt they needed that.

6.2.3 Data Analysis

6.2.3.1 Perceptual Data

Perceptual data were analyzed by determining the bias per condition for each participant. The bias is the difference between the Point of Subjective Equality (PSE) and the reference stiffness. This bias was obtained by first determining all the combinations of reference and test stimuli that were presented for that staircase pair. For each combination, the number of times that the participant responded that the test stimulus was the harder stimulus was counted and divided by the total number of trials in which the combination was presented, resulting in a response ratio. An example of such a data set is shown in Fig. 6.3b. To calculate the bias, a psychometric function was fitted to the data. For the condition in Experiment 2 in which the test stimulus had no damping and was therefore more lightly damped than the reference stimulus, the sign of the bias was reversed. This was done to make sure that the bias represented the difference between the more lightly and more heavily damped stimulus in a consistent manner. The following equation was used to describe the relation between test stiffness K_{test} and measured response ratio, with fitting parameters bias (μ) and JND (σ):

$$f(K_{test}) = \frac{1}{2} + \frac{1}{2}\mathrm{erf}\left(\frac{K_{test} - K_{ref} - \mu}{\sqrt{2}\sigma}\right). \tag{6.1}$$

To assess the goodness of fit to the measured data, the R^2 was calculated for each fit. When the R^2 was smaller than 0.25, the fit was deemed too poor and the bias was not used in further analyses. This was the case for 5 of the 72 calculated biases (7%) in Experiment 1, for 2 of the 110 calculated biases (2%) in Experiment 2, and for 2 of the 24 calculated biases (8%) in Experiment 3. In Experiment 2, one participant indicated at the debriefing that she had guessed the purpose of the experiment and therefore had manipulated her answers to make sure that she was 'not being fooled'. Therefore, her data were discarded from the analysis, resulting in a total of 11 participants in Experiment 2. The remaining biases were averaged over participants per task and reference stiffness. To assess if the biases differed significantly from 0, a Student's t-test was performed on the biases of each condition. For Experiment 1, the effect of reference stiffness was evaluated for both tasks separately, using a repeated measures ANOVA with reference stiffness as within-subject factor. For Experiment 2, the effect of damping was evaluated for both tasks separately, using a repeated measures ANOVA with damping as within-subject factor. When the sphericity-criterion was not met, Greenhouse-Geisser correction was used. For Experiment 3, the difference between the biases in both conditions was assessed using a paired t-test.

6.2.3.2 Position and Force Data

Position and force data were recorded to investigate if there were parameters that correlated to the perceptual experience of the participants, for instance, if participants always perceived a trial in which the force was higher as being the harder stimulus. Figure 6.4 shows an example of these data for one trial. If biases in the perceptual data are present, the position and force parameters could indicate which parameters participants used to base their percept on. Furthermore, a difference between tasks in the correlation of parameters and perceptual decision could indicate that participants switched strategies between tasks.

For each trial, the following characteristics were calculated:

Free-air phase:

- Movement time [s]
- Mean movement velocity [$\frac{m}{s}$]

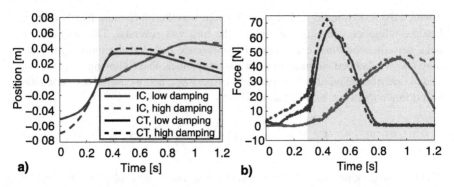

Fig. 6.4 Typical examples of (**a**) position data and (**b**) force data of one in-contact trial (gray, IC, stiffness for both movements 1000 N/m) and one contact-transition trial (black, CT, stiffness for both movements 2000 N/m). The solid lines show movements in the low-damped environment (5 Ns/m), while the dashed lines show movements in the high-damped environment (30 Ns/m). The object surface was placed at position 0, so the gray background shows the part of the movement in contact with the object. Note that for the in-contact task, the high-damped movement yields larger forces than the low-damped movement, while the displacement profile is similar. For the contact-transition task, initial forces are higher for the high-damped movement, but the force increase upon object contact is almost similar, while the indentation is larger for the high-damped movement

Impact phase:

- Impact velocity $[\frac{m}{s}]$
- Impact force [N]
- Extended rate-hardness $[\frac{N/s}{m/s}]$

In-contact phase:

- Movement time [s]
- Maximum object indentation [m]
- Mean movement velocity $[\frac{m}{s}]$
- Peak movement velocity $[\frac{m}{s}]$
- Mean deceleration during indentation $[\frac{m}{s^2}]$
- Peak deceleration during indentation $[\frac{m}{s^2}]$
- Mean force [N]
- Peak force [N]
- Force difference between impact and maximum object indentation [N]
- Adjusted rate-hardness $[\frac{N/s}{m/s}]$

Extended rate-hardness is defined as the *maximum rate of change of force* divided by the *movement velocity at impact*. This parameter was meaningless in our impact-free in-contact task, so we introduced the new parameter 'adjusted rate-hardness'. We calculated this by fitting straight lines to the force and position data as a function of time, between impact and maximum object indentation. The ratio of the two fitted slopes was called 'adjusted rate-hardness'. Note that the unit of adjusted and extended rate hardness, $[\frac{N/s}{m/s}]$, could be simplified to $[\frac{N}{m}]$. However, this was not done for extended rate-hardness in the original definition [6]. For consistency and to avoid confusion with a pure stiffness, the unit $[\frac{N/s}{m/s}]$ was used for both extended and adjusted rate-hardness. When no damping is present, rate-hardness and stiffness are the same, but for movements in which there is damping, a substantial additional force is present, such that rate-hardness and stiffness are no longer equivalent. The parameters for the in-contact phase were calculated for both tasks, while the parameters for the free-air phase and the impact phase were only used for the contact-transition task, as they were meaningless in an in-contact task. When a derivative was needed to calculate the parameter, the difference data were filtered using a fourth-order Butterworth low-pass filter with a 500 Hz cut-off frequency.

The calculated parameters were grouped based on the perceptual data: when participants answered that the first stimulus was the harder one, the parameter from the second trial was subtracted from that of the first and vice versa. When the difference score was positive, a '1' was scored, while for negative scores a '0' was noted. For each task, reference stiffness and participant, the scores of all trials were summed and divided by the total number of trials, resulting in a score between 0 and 1, which we called 'selection ratio'. The chance value would be 0.5, while a significant deviation from 0.5 would indicate a correlation between hardness perception and the measured parameter. For instance, a value above 0.5 would indicate that participants chose the stimulus that scored higher on this characteristic more often as the harder stimulus. Student's t-tests were used to assess if the parameters indeed deviated from 0.5. Only parameters that significantly differed from 0.5 for all conditions within an experiment are shown in the Results section, since they are correlated to the perceptual experience most strongly and consistently. The position and force data from participants and conditions in which the perceptual bias could not be determined reliably enough, as described in 'Perceptual data', were omitted from the analysis.

6.3 Results

6.3.1 Experiment 1

6.3.1.1 Perceptual Data

All the biases measured in Experiment 1 are shown in Fig. 6.5a. The in-contact task yielded negative biases, as shown in Table 6.1, while the contact-transition task yielded positive biases, as shown in Table 6.2. Almost all the biases differed significantly from 0 (Tables 6.1 and 6.2 show the exact values). The repeated measures ANOVAs showed a significant effect of reference stiffness on bias size in the contact-transition task ($F_{2,12} = 6.8$, $p = 0.011$), while there was no effect in the in-contact task ($F_{2,22} = 0.23$, $p = 0.80$). The different sign of the bias means that in the in-contact task, perceived hardness was increased when damping was added, while for the contact-transition task, perceived hardness was decreased

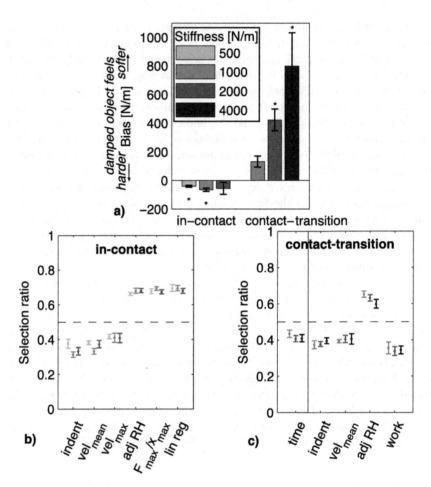

Fig. 6.5 Results of Experiment 1. (**a**) Biases for both tasks, averaged over participants. The error bars show ±1 standard error, while the gray values correspond to the various reference stiffnesses. A positive (negative) bias means that the heavily damped object is perceived as softer (harder) than the lightly damped object. Note that heavily damped objects were perceived as harder in the in-contact task, while they were perceived as softer in the contact-transition task. All biases differed significantly from 0, except for the bias of the in-contact task with a reference stiffness of 2000 N/m. $*p < 0.05$ when comparing bias to 0. (**b**) Parameters for in-contact task from movement and force data that correlated significantly with choices in the perceptual task, which means that they differed significantly from the chance level of 0.5 for all reference stiffnesses. When the ratio is larger (smaller) than 0.5, participants were inclined to connect an increase in this parameter to the object feeling harder (softer). Gray values correspond to the different reference stiffnesses, while error bars show ±1 standard error. (**c**) Significant parameters for contact-transition task. The vertical black line separates the parameters in the free-air phase from parameters in the in-contact phase. No parameters were significantly different from 0.5 for all stiffness levels in the impact phase

Table 6.1 Biases and statistics for the in-contact tasks in Experiments 1 and 2

Exp.	K [N/m]	B [Ns/m]	Bias $\pm$ s.e. [N/m]	t-value	p-value
1	500		-39 ± 6	$t_{11} = -6.4$	$\leq 0.001*$
	1000	30	-64 ± 12	$t_{11} = -5.4$	$\leq 0.001*$
	2000		-56 ± 42	$t_{11} = -1.3$	0.21
2		0	-20 ± 11	$t_{10} = -1.9$	0.092
		10	-17 ± 12	$t_{10} = -1.5$	0.17
	1000	15	-6.3 ± 12	$t_{10} = -0.51$	0.62
		20	-19 ± 9	$t_{10} = -2.0$	0.073
		25	-21 ± 12	$t_{10} = -1.7$	0.12

$* p < 0.05$ for a one-sample t-test with a test value of 0

Table 6.2 Biases and statistics for the contact-transition tasks in Experiments 1, 2 and 3

Exp.	K [N/m]	B [Ns/m]	Bias $\pm$ s.e. [N/m]	t-value	p-value
1	1000		131 ± 39	$t_{11} = 3.4$	0.006*
	2000	30	423 ± 76	$t_{10} = 5.6$	$\leq 0.001*$
	4000		798 ± 234	$t_7 = 3.4$	0.011*
2		0	20 ± 41	$t_{10} = 0.48$	0.64
		10	100 ± 27	$t_{10} = 3.7$	0.004*
	2000	15	146 ± 72	$t_{10} = 2.0$	0.071
		20	149 ± 70	$t_9 = 2.1$	0.062
		25	223 ± 58	$t_9 = 3.9$	0.004*
3	2000	test global	400 ± 130	$t_{10} = 3.1$	0.012*
		test local	-170 ± 80	$t_{10} = -2.3$	0.046*

$* p < 0.05$ for a one-sample t-test with a test value of 0

when damping was added. Since almost all the biases differed significantly from 0, this implies that the biases also differ significantly between the tasks. The results of Experiment 1 have been published previously in conference proceedings [21].

6.3.1.2 Position and Force Data

The position and force data of Experiment 1 (see Fig. 6.5b, c) showed 4 parameters that significantly differed from chance for all of the reference stiffnesses in the in-contact task, which were: maximum indentation, mean velocity, maximum velocity and adjusted rate-hardness (all $p \leq 0.0068$). For the contact-transition task, the following 4 parameters differed significantly from chance for all the reference stiffnesses: movement time during the free-air phase, maximum object indentation, mean velocity and adjusted rate-hardness (all $p \leq 0.037$). The latter 3 parameters concerned the in-contact phase of the movement.

6.3.2   Experiment 2

6.3.2.1   Perceptual Data

All the biases measured in Experiment 2 are shown in Fig. 6.6a. The in-contact task yielded negative biases, as shown in Table 6.1, while the contact-transition task yielded positive biases, as shown in Table 6.2. For the in-contact task, none of the biases differed significantly from 0, while some biases did differ from 0 in the contact-transition task (Tables 6.1 and 6.2 show the exact values). The repeated measures ANOVAs showed a significant effect of damping on bias size in the contact-transition task ($F_{2,12} = 6.8$, $p = 0.011$), while there was no effect in the in-contact task ($F_{2,12} = 0.23$, $p = 0.80$). Similar to Experiment 1, biases for the in-contact task were negative, while biases for the contact-transition task were positive, implying a difference between the tasks.

6.3.2.2   Position and Force Data

The position and force data of Experiment 2 (see Fig. 6.6b, c) showed 4 parameters that differed significantly from chance for all conditions in the in-contact task, which were: maximum indentation, mean velocity, maximum velocity and adjusted rate-hardness (all $p \leq 0.0093$). For the contact-transition task, the following 3 parameters differed significantly from chance for all conditions: maximum object indentation, mean velocity and adjusted rate-hardness (all $p \leq 0.037$). All the parameters concerned the in-contact phase of the movement.

6.3.3   Equal Hardness Lines

The aim of Experiment 1 was to investigate the influence of different levels of stiffness on perceived hardness, while Experiment 2 had the same aim for different levels of damping. In both experiments, the two different tasks were used. To combine all the information obtained in the two experiments, Fig. 6.7 was created. In the left panel, the information for the in-contact task in both experiments is summarized, while in the right panel, the data from the contact-transition task are shown. Along the vertical direction, the influence of reference stiffness can be seen, while along the horizontal direction, the influence of damping is shown. Note that the dotted lines only connect equal hardness pairs, they are not a suggestions of the shape of the connecting lines. The solid lines do indicate the effect of damping on perceived hardness, as the data of Experiments 1 and 2 together do provide enough data to discern a relation between damping and perceived hardness for the reference stiffness of 1000 N/m for the in-contact task and for the reference stiffness of 2000 N/m for the contact-transition task. For the in-contact task, the shape of the data most closely resembled a linear relation, so a linear function was fitted to the

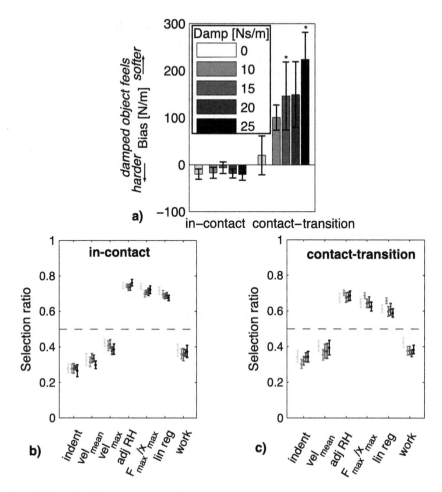

Fig. 6.6 Results of Experiment 2. (**a**) Biases for both tasks, averaged over participants. The error bars show ±1 standard error, while the gray values correspond to the various levels of damping of the test stimulus. For the in-contact task, the reference stiffness was always 1000 N/m, while it was always 2000 N/m for the contact-transition task. Note that heavily damped objects were perceived as harder in the in-contact task, while they were perceived as softer in the contact-transition task. The biases for 10 and 25 Ns/m differed significantly from 0 in the contact-transition task. *$p < 0.05$ when comparing biases to 0. (**b**) Parameters for in-contact task from movement and force data that correlate significantly with choices in the perceptual task. Gray values correspond to the different damping levels, while error bars show ±1 standard error. (**c**) Significant parameters for contact-transition task. No parameters were significantly different from 0.5 for all damping levels in the free-air phase or the impact phase. See Fig. 6.5 for a more elaborate caption

data to describe an equal hardness line, resulting in an R^2 of 0.78. For the contact-transition task, the shape most closely resembled a power function, which was fitted to that part of the data to describe the equal hardness line, resulting in an R^2 of 0.91. A linear fit to the latter data resulted in an R^2 of 0.89, so fitting a power function

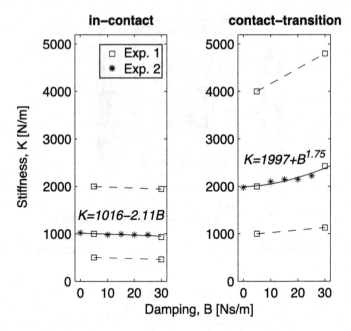

Fig. 6.7 Equal hardness lines, for the in-contact task (left) and the contact-transition task (right), from Experiments 1 and 2. Note that the dotted lines only connect stiffness-damping pairs with equal perceptual hardness, but they are not intended as an interpolation between the two pairs. The fitted solid lines do show an interpolation, based on a combination of some data of Experiment 1 and all data of Experiment 2. For the in-contact task, the increase in perceived hardness (stronger negative bias), can be described using a linear fit. For the contact-transition task, a power function is more appropriate to describe the decrease in perceived hardness (stronger positive bias)

was indeed slightly more appropriate for the contact-transition task. So, the panels of Fig. 6.7 provide an image of pairs of damping and stiffness values which resulted in an equal sensation of hardness in the two different tasks in these experiments.

6.3.4   Experiment 3

6.3.4.1   Perceptual Data

Figure 6.8a shows an overview of the biases that were measured in the two inter-leaved conditions in Experiment 3. For the condition in which the test stimuli were globally damped, while the reference stimuli were locally damped, the measured bias was positive. For the condition in which the test stimuli were locally damped, while the reference stimuli were globally damped, the measured bias was negative. Both biases differed significantly from 0 (see Table 6.2), which implies that the

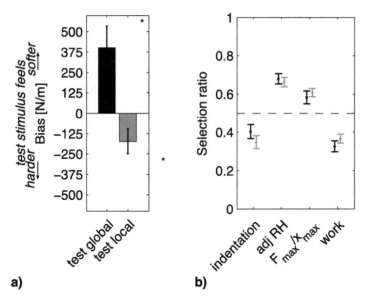

Fig. 6.8 Results of Experiment 3. (**a**) Biases for both conditions, averaged over participants. The error bars show ±1 standard error. The labels on the horizontal axis indicate the type of damping of the test stimulus. The biases had a different direction, and differed significantly from 0. This indicates the consistency of the effect: when the test stimulus is globally damped, it feels softer than the locally damped reference stimulus, while when the test stimulus is locally damped, it feels harder than the globally damped reference stimulus. *$p < 0.05$ when comparing biases to 0. (**b**) Parameters from movement and force data that correlated significantly with choices in the perceptual task. Gray values correspond to the different damping conditions, while the error bars represent ±1 standard error. See Fig. 6.5 for a more elaborate caption

biases also differ from each other. This shows the consistency of the effect: when the test stimulus was globally damped, it was perceived as softer than the locally damped reference stimulus, while when the test stimulus was locally damped, it was perceived as harder than the globally damped reference stimulus. However, a paired t-test showed that the magnitude of the biases did differ significantly ($t_{10} = 2.9$, $p = 0.017$).

6.3.4.2 Position and Force Data

Figure 6.8b shows that only 2 parameters in the force and position data differed significantly from chance for both conditions in Experiment 3: maximum indentation and adjusted rate-hardness (all $p \leq 0.033$). Both significant parameters concerned the in-contact phase of the movement. During the free-air movement and the impact phase, none of the parameters differed significantly from 0 in both conditions.

6.4   Discussion

In this study, a large effect of damping on perceived hardness was found. Interestingly, this effect was very different for the two experimental tasks: adding damping *increases* the perceived hardness for an in-contact task, while it *decreases* it for a contact-transition task. This effect was found in both Experiment 1 and Experiment 2. Experiment 3 indicated that this effect was much larger for global damping (present throughout the environment) than for local damping (present inside the object only).

Apart from the direction of the effect being task-dependent, the size of the effect is also very different between the tasks: the decrease for the contact-transition task was much larger than the increase for the in-contact task, as shown in Experiment 1. The highest damping level resulted in a relative bias of about ∼20% for the contact-transition task, which is close to the JND of about ∼23% for stiffness [9]. This means that this stiffness difference would be (close to) perceivable in an undamped situation. In Experiment 2, the same difference in effect size between the tasks was found. For this experiment, the decrease for the in-contact task was not significantly different from 0 for any of the damping levels, while it was positive for the contact-transition task. Apparently, a damping level of 30 Ns/m was needed to produce a significant bias in the in-contact task, while a damping level of 10 Ns/m was enough to produce a significant bias in a contact-transition task. So, while Lawrence et al. [12] argue that dampers could be used to provide a haptic illusion of hard objects, our results show that this is very task-dependent and could even decrease perceived hardness in a contact-transition situation. Contact-transition situations occur frequently in bilateral teleoperation [16], so for these applications, adding damping should be done carefully.

To understand where the perceptual differences between the tasks come from, the force and position data can provide relevant information, as they can give insight into the parameters on which participants based their perceptual decision. In all experiments and in all conditions, maximum object indentation and adjusted rate-hardness were significantly correlated with the perceptual experience of hardness. Mean velocity was significantly correlated in all conditions of Experiments 1 and 2, but not in Experiment 3. The direction of the correlation was always the same, so a smaller maximum indentation, a smaller mean velocity and a larger adjusted rate-hardness were correlated with an object feeling harder. Interestingly, in almost all conditions, only parameters describing the in-contact part of the movement were significantly correlated to the perceptual decision. We had expected parameters describing the impact phase of the movement to be important in the contact-transition task, such as impact force or extended rate hardness, as it is known that contact transients can be very important for simulating hard surfaces [10]. Apparently, our participants followed the instruction to focus on the in-contact part of the movement. Moreover, in both tasks almost the same parameters came out as significant in the same direction, even though the direction of the perceptual effect was reversed between the two tasks. So, it seems that participants did not use different strategies in the different tasks, but there was an actual difference in the task dynamics. Intuitively, this does make sense. When adding damping in an in-contact task, an extra force is added during indentation. So, if a participant would keep his force profile constant, the damping would decrease the amount of object indentation, making the object feel harder. For a contact-transition task, adding damping would make the transition from free-air motion to object indentation a bit more gradual, which would allow the participant to indent the damped object a bit further when using the same force profile, making the object feel softer. The notion that no force parameters were significantly correlated to the perceptual experience strengthens the hypothesis that participants tried to keep their force profile constant in both parts of the trial, while they assessed the kinematic response of the system.

To get a grasp on the interplay between stiffness and damping in hardness perception, we constructed the 'equal hardness lines'. These lines illustrate that for the contact-transition task, increasing the level of damping or increasing the reference stiffness increased the bias in perceived hardness. For the in-contact task, no significant effect of reference stiffness or damping level on perceptual bias was found. However, when performing a linear fit on the biases from the in-contact task, shown in the left graph of Fig. 6.7, the slope is negative, which suggests that increasing the level of damping does increase the bias. However, all biases were very small in this task, and only the maximum level of damping created a significant bias. The two panels together can be seen as a start of a guideline, which indicates the effect of injecting damping in a system on the object hardness perceived by the user of the system. However, the extreme task-dependency of the effect, shown by the reverse of the bias for the different tasks, already indicates that it is virtually impossible to provide one guideline to fit all situations. Moreover, the size of the error bars in, for instance, Fig. 6.6 also indicates that the size of the bias was quite different among participants, even though the direction of the effect was very consistent. Finally, it is important to note that in natural object manipulation, local surface deformation is a very important cue for object hardness [2, 18]. So, our results cannot be extended directly to a natural situation. However, in haptic devices, which are often being used in teleoperation systems, there is usually no sensory information about local surface deformation, as the rigid interface only provides global force and movement information.

It is striking that existing literature describes an increase in perceived hardness when adding damping to a system, even though tasks were used that are comparable to our contact-transition task [6, 12]. However, both Lawrence et al. and Han and Choi used a form of local damping in their experiments. Our Experiment 3 clearly shows a large perceptual difference between adding global and local damping: for the globally damped test stimuli we found a positive bias, while we found a negative bias for the locally damped test stimuli. These results are congruent with the positive biases for the in-contact task in Experiments 1 and 2, since the test stimuli were also damped globally there. The inversion of the bias with inversion of the reference stimulus shows the consistency of the effect: irrespective of which stimulus was deemed test stimulus, the globally damped stimulus was perceived as softer than the locally damped one. The magnitude of the bias for the locally damped test stimuli was smaller than that for globally damped test stimuli, but the mean stiffness during the experiment was also smaller for the former condition, because the reference stiffness was the same in both conditions. Since a staircase converges onto the bias, the mean stiffness of the stimuli depends on the measured bias. Experiment 1 already showed that the size of the bias depends on the stiffness of the objects, so a smaller mean stiffness should result in a smaller bias. Concluding, it seems that, even though other factors like device inertia, maximum force and amount of indentation could still play a role, the difference between global and local damping is at least an important part of the explanation of the difference between previous results and our results. Especially for the application in teleoperation systems, results on global damping are very relevant, since the addition of damping is often used to increase system stability [see e.g. 8, 14, 19].

6.5   Conclusion

We found that adding global damping to a system *increases* the perceived hardness of an object composed of a spring in an in-contact task, while it *decreases* the perceived hardness in a contact-transition task. The latter effect is larger than the former. Participants seem to have used the same strategy in both tasks: using the same force-profile during the two movements of one trial and assessing the reaction of the system to this force. Object indentation, mean velocity and adjusted rate-hardness were closely related to the perceptual experience of hardness. This knowledge could be used by designers of teleoperation systems, as they can assess the effect of design choices on the reflected object hardness by their controllers, given the task and required amount of injected damping. Our 'equal hardness lines' could be used as a preliminary guideline in this process. Concluding, our results clearly show that using a damper to increase the perceived hardness of virtual objects is only effective in tasks without impact, while it is detrimental in contact-transition tasks.

References

1. Bergmann Tiest WM (2010) Tactual perception of material properties. Vis Res 50(24):2775–2782
2. Bergmann Tiest WM, Kappers AML (2009) Cues for haptic perception of compliance. IEEE Trans Haptic 2(4):189–199
3. Blair GWS, Coppen FMV (1939) The subjective judgment of the elastic and plastic properties of soft bodies; the "differential thresholds" for viscosities and compression moduli. Proc R Soc Lond Ser B Biol Sci 128(850):109–125
4. Coppen F (1942) The differential threshold for the subjective judgement of the elastic and plastic properties of soft bodies. Br J Psychol Gen Sect 32(3):231–247
5. Coren S (1993) The left-hander syndrome. Vintage Books, New York

6. Han G, Choi S (2010) Extended rate-hardness: a measure for perceived hardness. In: Kappers AML, van Erp JBF, Bergmann Tiest WM, van der Helm FCT (eds) Haptics: generating and perceiving tangible sensations. Lecture notes in computer science, vol 6191. Springer, Berlin/Heidelberg, pp 117–124

7. Harper R, Stevens SS (1964) Subjective hardness of compliant materials. Q J Exp Psychol 16(3):204–215

8. Hokayem PF, Spong MW (2006) Bilateral teleoperation: an historical survey. Automatica 42:2035–2057

9. Jones LA, Hunter IW (1990) A perceptual analysis of stiffness. Exp Brain Res 79(1):150–156

10. Kuchenbecker K, Fiene J, Niemeyer G (2006) Improving contact realism through event-based haptic feedback. IEEE Trans Vis Comput Graph 12(2):219–230

11. Lawrence D (1993) Stability and transparency in bilateral teleoperation. IEEE Trans Robot Autom 9(5):624–637

12. Lawrence D, Pao LY, Dougherty A, Salada M, Pavlou Y (2000) Rate-hardness: a new performance metric for haptic interfaces. IEEE Trans Robot Autom 16(4):357–371

13. Nuño E, Basañez L, Ortega R, Spong MW (2009) Position tracking for non-linear teleoperators with variable time delay. Int J Robot Res 28:895–910

14. Nuño E, Basañez L, Ortega R (2011) Passivity-based control for bilateral teleoperation: a tutorial. Automatica 47:485–495

15. Rosenberg L, Adelstein B (1993) Perceptual decomposition of virtual haptic surfaces. In: Proceedings of the IEEE symposium on research frontiers in virtual reality, San Jose, pp 46–53

16. Sarkar N, Yun X (1996) Design of a continuous controller for contact transition task based on impulsive constraint analysis. In: Proceedings of the 1996 IEEE international conference on robotics and automation, vol 3, pp 2000–2005

17. Sheridan T (1993) Space teleoperation through time delay: review and prognosis. IEEE Trans Robot Autom 9(5):592–606

18. Srinivasan MA, LaMotte RH (1995) Tactual discrimination of softness. J Neurophysiol 73(1):88–101

19. Sun D, Naghdy F, Du H (2014) Application of wave-variable control to bilateral teleoperation systems: a survey. Annu Rev Control 38:12–31

20. Van der Linde R, Lammertse P (2003) HapticMaster – a generic force controlled robot for human interaction. Ind Robot Int J 30(6):515–524

21. Van Beek FE, Heck DJF, Nijmeijer H, Bergmann Tiest WM, Kappers AML (2015) The effect of damping on the perception of hardness. In: Proceedings of the IEEE world haptics conference (WHC), Evanston, pp 82–87

Part III
Applications

Chapter 7
Integrating Force and Position

Abstract In this study, we investigated the integration of force and position information in a task in which participants were asked to estimate the center of a weak force field. Two hypotheses, describing how participants solved this task, were tested: (1) by only using the position(s) where the force reaches the detection threshold, and (2) by extrapolating the force field based on perceived stiffness. Both hypotheses were also described formally, assuming a psychophysical function obeying a power law with an exponent smaller than one. The hypotheses were tested in two psychophysical experiments, in which 12 participants took part. In Experiment 1, an asymmetric force field was used and the presence of visual feedback about hand position was varied. In Experiment 2, a uni-lateral force field was used. For both experiments, hypothesis 1 predicts biases between (Experiment 1) or at (Experiment 2) the position(s) of the force detection threshold, while hypothesis 2 predicts smaller biases. The measured data show significant biases in both experiments that coincide with the biases predicted by using force detection thresholds from literature. The average measured responses and their variabilities also fitted very well with the mathematical model of hypothesis 1. These results underline the validity of hypothesis 1. So, participants did not use a percept of the stiffness of the force field, but based their estimation of the center of the force field on the position(s) where the force reached the detection threshold. This shows that force and position information were not integrated in this task.

Previously published as:
F.E. van Beek, W.M. Bergmann Tiest, A.M.L. Kappers & G. Baud-Bovy
Integrating force and position: testing model predictions
Experimental Brain Research 234(11), 3367–3379.

© Springer International Publishing AG 2017   105
F.E. van Beek, *Making Sense of Haptics*, Springer Series on Touch and Haptic
Systems, https://doi.org/10.1007/978-3-319-69920-2_7

7.1 Introduction

Even though we perceive the world through many different sensory modalities, which provide us with different types of information about the world, we still have only one percept of the world. Recently, there has been a lot of attention in the perceptual literature on how we deal with all these different sources of information. A very influential idea is the maximum-likelihood estimation model, which describes the integration of different sensory estimates as a statistically optimal integration, influenced by prior knowledge of the world [7, 23]. This idea can be applied to the integration of information across modalities [8] or across different types of information within the same modality [12, 16]. In our study, we investigated the integration of information within the haptic modality.

A striking example of the combination of haptic cues is a study by Robles-De-La-Torre [18], in which participants were asked to discriminate between holes and bumps. Even if the geometrical properties signaled a bump, a change in the force cues could cause the bumps to be perceived as holes. In a comparable experimental paradigm, Drewing and Ernst [6] showed that weights of the force and position cues in curvature perception depend on the type of shape that is being explored: when the shape is strongly curved, position cues are weighted more strongly, while force cues are weighted more strongly when the shapes are flatter. In our study, we were also interested in the integration of force and position cues within the haptic modality. Specifically, we investigated the integration of force and position information in the haptic perception of the centre of a weak elastic force field.

This question is particularly interesting when the force field of interest is very weak, as the forces already become imperceptible at some distance from the true centre of the force field. The threshold for correctly discriminating between a right- and leftward force is 0.1 N in a static situation and 0.05 N when movement is allowed [3], so as soon as forces exceed these values at some distance from the centre only, humans somehow have to infer the true centre of the force field. By studying these weak force fields, we can learn something about how humans combine position and force information over time, as they have to acquire the information by exploring the force field. In this study, we tested two hypotheses, which were: (1) participants do not use stiffness information, but only use the position at which the force becomes imperceptible, which we call the bisection model; and (2) participants estimate the stiffness of the force field through exploration. By extrapolating the forces based on the estimated stiffness, they estimate the position of the centre of the force field, which we call the stiffness model. The former model has been described previously [2, 4].

If hypothesis 2 is correct, participants use stiffness information, which they gathered during the exploration of the force field. It is not known yet which parameters we use to perceive stiffness. Jones and Hunter [13] found the discrimination threshold of stiffness to be higher than those of force and position, but they did show that participants used both force and position cues in their stiffness discrimination task. Tan et al. [22] underlined the relevance of work cues, while Srinivasan and

La Motte [19] suggested that the rate of change of average pressure might play an important role. There is also a considerable amount of work on the integration of visual and haptic information in stiffness perception [e.g. 14, 15]. Irrespective of what parameters are being used to estimate stiffness, all theories share the fact that information needs to be acquired in a serial fashion, since stiffness is meaningless in a static situation. So, hypothesis 2 implies that participants acquire information over time by exploring the force field. By doing this, they build a percept of the stiffness of the force field and use the product of all this information to extrapolate the center of the force field. Importantly, this hypothesis also implies that position and force information need to be integrated, in order to obtain an estimate of the stiffness. So, if this hypothesis is true, integration between force and position information takes place in this task.

If hypothesis 1 is correct, the nature of the task is less sequential than in the case of hypothesis 2. Hypothesis 1 assumes that by exploring the force field, participants find the position(s) where the force just reaches the detection threshold. In the case of a bi-lateral force field (present in Experiment 1), they assume the center of the force field to be in between the two positions where the force reaches the threshold, while for a uni-lateral force field (present in Experiment 2), they assume the center of the force field to be at the position where the force reaches the threshold. If this is true, knowing the force threshold and the stiffness of the force field is enough to predict the position which participants will perceive to be the center of the force field. So, if hypothesis 1 is true, participants do not integrate force and position information to obtain a percept of the stiffness in this task, even though they do explore the force field and thus could have access to this information.

We designed two experiments to test the validity of the hypotheses. In Experiment 1, an asymmetric force field was used. In Experiment 2, a uni-lateral force field was used. For both experiments, hypothesis 1 (the bisection model) predicts biases between (Experiment 1) and at (Experiment 2) the force detection threshold, while hypothesis 2 (the stiffness model) predicts smaller biases. Moreover, in Experiment 1 the visual feedback was manipulated, which was predicted to influence the noise in the responses. Both experiments can differentiate between the two hypotheses and can thus contribute to testing if the brain integrates force and position information in this task.

7.2 Models

In this section, the mathematical descriptions of the bisection model and the stiffness model are explained. Both models are used to describe the behavior of participants who are asked to find the center of a force field in a one-dimensional situation (i.e. along a line). Two situations are described in particular: (1) a bilateral force field, in which one spring is active in the region left of the central position of the force field and the other is active in the region to the right of the center of the force field. These springs can be different, creating an asymmetric force field; and (2) a unilateral force field, in which only one spring is present. In this case, the term 'center of the force field' is used to indicate the point where the spring is attached, so the position along the line where the force just becomes 0. Both models can predict the bias in finding the center of a force field. These values can also be determined experimentally, which allows for a comparison between the validity of the two models based on the experimental data. In Fig. 7.1, a visual explanation of the predictions of these models for finding the center of a force field is given. For both models, a stationary reference frame is used, of which the origin is positioned at the center of the force field. To be able to predict the biases, the relation between the physical and estimated (or perceived) force magnitude needs to be known. We assume that the physical (F) and perceived ($\tilde{F}$) force are related through a power function [20, 21]:

$$\tilde{F} = \alpha(F - F_{\text{th}})^{\beta}. \qquad (7.1)$$

In this equation, F_{th} is the force threshold above which the relation exists, while α and β determine the the the shape of the relation. In our previous experiments, we measured force perception for a range of forces using a haptic device. In those experiments, we found a mean β of 0.8 [24], so we will assume this β for the current experiment too. In a linear force field, the force is dependent on the spring constant (K) and the position (x), so we can describe Eq. 7.1 in the following way in this situation:

$$\tilde{F} = \alpha(K(x - x_0))^{\beta} \qquad (7.2)$$

with x_0 being the position where the perceived force is 0. Note that for Experiment 2, a minus sign needs to be added before K, since the force field is an attractive one in this experiment. Starting from Eq. 7.2, participants could use 2 approaches to find the center of the force fields: by only using the position(s) where the perceived force just reaches 0 (i.e. the bisection model, which is hypothesis 1), or by probing the relation between position and force and extrapolating the position were the force becomes 0 (i.e. the stiffness model, which is hypothesis 2). The models describing the two hypotheses are explained in more detail below.

7.2.1   Bisection Model

The bisection model was first proposed in Bocca and Baud-Bovy [4] while an extended version is described in detail in Baud-Bovy [2]. The model is based on the idea that the participant does not use stiffness information, but only uses the position(s) where the force reaches the threshold. In the model, parameters describing the force threshold and the perception of hand position are used. Furthermore, a weight factor between the left and right positions where the force

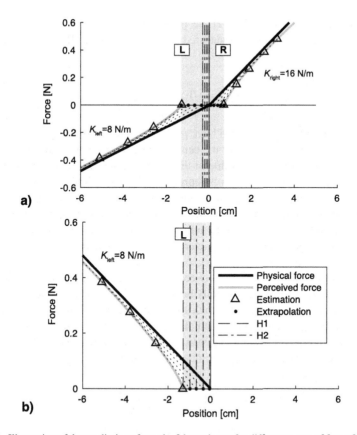

Fig. 7.1 Illustration of the predictions from the 2 hypotheses for different types of force fields. The vertical gray area indicates the area of the workspace in which the force is below threshold level and thus imperceptible to the participant. The width of this area depends on the spring stiffness (K) of the force field, which is indicated with the black line. (**a**) For an asymmetric bilateral force field, the bisection model predicts a shift of the end position towards the weaker force field (dashed line, H1), in the middle of the right (R) and the left (L) positions of the force threshold. The stiffness model predicts a bias in the same direction, but a weaker one. For this figure, it is assumed that the stiffness estimates at both sides of the force field are made at the same force magnitude (the triangles show the positions and forces at which the estimates are made), from which the edges of the force fields are extrapolated (black dots). The middle between the black dots is the prediction of the estimated center of the force field from the stiffness model (dashed-dotted line, H2). This results in a decrease in bias with an increase of position at which the stiffness estimate is made, up to biases disappearing for positions at the edge of the workspace. For both hypotheses, the direction of the predicted bias does not depend on the exact values of the spring stiffnesses (K_{left} and K_{right}), but only on which spring is weaker. (**b**) In a unilateral force field, only one spring is present, so the right half of the figure is empty. The bisection model predicts an end position at the force threshold (dashed line, H1). It also predicts larger biases for weaker springs, since the vertical gray area is larger in that case. The stiffness estimation hypothesis predicts weaker biases (dashed-dotted line, H2), up to biases disappearing when stiffness is estimated using the largest forces in the force field

reaches the threshold is added. These parameters are combined to describe the bias and the variability in the measured data. For the bias, Baud-Bovy [2] used the following:

$$\mu_x = (1 - w)\frac{-F_{\text{th}}}{K_{\text{left}}} + w\frac{F_{\text{th}}}{K_{\text{right}}}. \qquad (7.3)$$

So the estimation of the measured bias, μ_x, is based on the mean position between the two positions left and right of the center where the force reaches the threshold. These positions depend on the value of the force threshold, F_{th}, and the stiffness of the force field that is used, which is represented by the stiffness of the left spring, K_{left}, and the right spring K_{right}. Note that K in this equation is a value that is only used to be able to make predictions, it is not a parameter that participants need to estimate, since they only need to estimate their hand position. In contrast, the value of $\tilde{K}$ in the stiffness model is a parameter of which participants do need to make an estimate.

The factor w is a weight factor between the two positions, which was introduced in the model in Baud-Bovy [2], mainly because the device was not positioned at the body midline in their experiment, which appeared to cause a shift in their data. In our experiment, the device was positioned at the body midline, so there was no reason to expect a difference in weight between the left and the right positions. Therefore, w was fixed at $\frac{1}{2}$ in the current experiment, resulting in:

$$\mu_x = \frac{1}{2}\left(\frac{-F_{\text{th}}}{K_{\text{left}}} + \frac{F_{\text{th}}}{K_{\text{right}}}\right). \qquad (7.4)$$

To describe the within-participant variability between the trials, the following equation was used:

$$\sigma_x^2 = (1 - w)^2\left(\frac{\sigma_{F_{\text{th}}}^2}{K_{\text{left}}^2} + \sigma_P^2\right) + w^2\left(\frac{\sigma_{F_{\text{th}}}^2}{K_{\text{right}}^2} + \sigma_P^2\right). \qquad (7.5)$$

The within-participant variability is σ_x, while $\sigma_{F_{\text{th}}}$ and σ_P describe the variability in the force threshold and the perception of hand position, respectively. The assumption in the model is that the variability in both signals is combined optimally, which is done mathematically by adding the variances, resulting in a final prediction of the measured variability. Again, w was fixed at $\frac{1}{2}$ in the current experiment, resulting in:

$$\sigma_x^2 = \frac{1}{4}\left(\frac{\sigma_{F_{\text{th}}}^2}{K_{\text{left}}^2} + \frac{\sigma_{F_{\text{th}}}^2}{K_{\text{right}}^2} + 2\sigma_P^2\right). \qquad (7.6)$$

In one of our experiments, we used two types of visual feedback about hand position to assess its effect on the positional noise parameter. To be able to fit the data of both of the visual feedback conditions together to make sure that as few parameters as possible were used, Eq. 7.6 was modified to:

$$\sigma_x^2 = \frac{1}{4}\left(\frac{\sigma_{F_{th}}^2}{K_{left}^2} + \frac{\sigma_{F_{th}}^2}{K_{right}^2} + 2k\sigma_{P_{fp}}^2 + (1-k)2\sigma_{P_{fa}}^2\right). \qquad (7.7)$$

We assumed the visual feedback to only affect the positional noise parameter, since that was the only parameter that the visual feedback provided any information about. Therefore, we included two position parameters, while fitting the same bias, force threshold and noise of the force threshold for all data. The positional noise for the condition in which visual feedback was present was $\sigma_{P_{fp}}$, while the noise for the condition in which visual feedback was absent was $\sigma_{P_{fa}}$. The value of k was used as a switch: for the condition in which visual feedback was present, k was 1 and for the condition without visual feedback, it was 0.

If the force field is unilateral, so only one spring is present, the bisection model can be simplified to:

$$\mu_x = \frac{F_{th}}{K} \qquad (7.8)$$

for the bias, and

$$\sigma_x^2 = \frac{\sigma_{F2}}{K^2} + \sigma_P^2 \qquad (7.9)$$

for the variability. In the fitting procedure, a least-squares approximation was used, in which the Sum of Squares (SS) was minimized for the bias and variability together:

$$SS = \sum_{i=1}^{n}(x_i - \mu_{x_i})^2 + (s_i - \sigma_{x_i})^2. \qquad (7.10)$$

In this equation, i refers to the number of the condition, while n is the total number of conditions. The measured bias in the i-th condition is x_i, while the measured variability is s_i. The bias and variability derived from the model for that condition are μ_{x_i} and σ_{x_i}, respectively. This fitting procedure can be used for both experiments and both models.

7.2.2 Stiffness Model

In the stiffness model, it is assumed that participants estimate the stiffness of the force field by exploring it. When they have formed this estimate, they can use it to find the distance from their current position to the position where the force becomes 0 by extrapolating the current force, as long as it is above the force threshold, through:

$$\tilde{x}_d = -\frac{\tilde{F}}{\tilde{K}}. \tag{7.11}$$

In this equation, $\tilde{x}_d$ is a signed quantity that indicates the movement needed to reach the edge of the force field, while $\tilde{F}$ is the force sensed at the current hand position, and $\tilde{K}$ is the estimated stiffness. Note that for Experiment 2, the force field is reversed, so to describe that situation, the minus sign before $\tilde{F}$ needs to be removed. When combining this equation with Eq. 7.2, we obtain:

$$\tilde{x}_d = -\frac{\alpha(K(x - x_0))^\beta}{\tilde{K}}. \tag{7.12}$$

We assume that the estimated stiffness corresponds to the slope of the psychophysical function:

$$\tilde{K} = \frac{d\tilde{F}}{d\tilde{x}} = \frac{d\tilde{F}}{dx}\frac{dx}{d\tilde{x}}. \tag{7.13}$$

Since we assume the position signal to be unbiased, the last term equals 1 and thus disappears. By now combining Eqs. 7.2 and 7.13, we obtain:

$$\tilde{K} = \frac{d}{dx}\alpha(K(x - x_0))^\beta = K\beta\alpha(K(x - x_0))^{\beta-1}. \tag{7.14}$$

By now combining Eqs. 7.12 and 7.14, we obtain a new description of $\tilde{x}_d$:

$$\tilde{x}_d = -\frac{\alpha(K(x - x_0))^\beta}{K\beta\alpha(K(x - x_0))^{\beta-1}} = -\frac{(x - x_0)}{\beta}. \tag{7.15}$$

If the position (x) of the handle changes, the perceived force $(\tilde{F})$ changes and thus $\tilde{x}_d$ changes. This allows for multiple estimates $(\tilde{x}_c)$ of the edge of the same force field, through:

$$\tilde{x}_c = x + \tilde{x}_d = x - \frac{(x - x_0)}{\beta}. \tag{7.16}$$

So, in this model, the estimated center of the force field depends on the current position, the position at which the perceived force becomes 0, and β. The final bias in a bilateral force field must be based on an integration of the extrapolations at the 2 sides, through:

$$\tilde{x}_c = \frac{1}{2}(\tilde{x}_{c_\text{left}} + \tilde{x}_{c_\text{right}}). \tag{7.17}$$

By comparing the perceived and the actual position of the center of the force field, the bias can be obtained:

$$\mu_x = \tilde{x}_c - x_c. \tag{7.18}$$

As the actual position of the center of the force field is placed at position 0, Eqs. 7.16, 7.17, and 7.18 can be combined to describe the bias:

$$\mu_x = \frac{1}{2}\left((x_{\text{left}} - \frac{x_{\text{left}} - x_{0_{\text{left}}}}{\beta}) + (x_{\text{right}} - \frac{x_{\text{right}} - x_{0_{\text{right}}}}{\beta}) \right) \tag{7.19}$$

which can be simplified to:

$$\mu_x = \frac{1}{2}\left(x_{\text{left}} + x_{\text{right}} - \frac{x_{\text{left}} + x_{\text{right}} - x_{0_{\text{left}}} - x_{0_{\text{right}}}}{\beta} \right). \tag{7.20}$$

To qualitatively describe this situation, Fig. 7.1 was made. From this figure, it is clear that the estimation of the position of the center of the force field in the stiffness model is closer to the true center of the force field than in the bisection model, for both experiments. However, the precise position of the estimation depends on the hand position at which the estimation is made in the stiffness model, which is unknown. So, we can only obtain a quite large range of predictions, and not one value. Therefore, our approach will be as follows: we will first predict the biases using the bisection model, while taking force detection thresholds from literature (0.05–0.1 N, as reported in Baud-Bovy and Gatti [3]). If the measured biases are smaller than the predicted ones, this suggests the stiffness model describes the data more appropriately. In that case, we will make additional assumptions about hand position to be able to fit the stiffness model, which will also allow us to make predictions about the variability of the data. If we do obtain force thresholds that fit the predicted values, the stiffness model cannot explain the data, and we proceed with fitting the bisection model only. In the latter situation, the bisection model will be fitted to the data with the force detection threshold being a free parameter, so the fitted detection threshold is also expected to lie between 0.05 and 0.1 N.

7.3   Material and Methods

7.3.1   Participants

In Experiment 1, 12 participants took part, 2 females and 10 males. They were 31 ± 6 years old; 10 were right-handed and 2 were left-handed. In Experiment 2, 12 other participants took part, 3 females and 9 males. They were 34 ± 14 years old, and all were right-handed. None of the participants had a history of neurological disorders. All participants signed an informed consent form and were given written

Fig. 7.2 Participant holding
the handle of the Omega.3.
The device is restricted to
movements along a line
parallel to the frontal axis of
the participant, so participants
can only make left-right
movements. The response
box, which is normally held
in the non-dominant hand, is
lying on the table in the lower
right corner of the picture. On
the screen, the feedback
presented in Experiment 1 is
shown: the position of the dot
on the line indicates the
participant's hand position,
while the color of the dot
indicates if the grip force is
within the required range

instructions prior to the experiment, while being naive to the purpose of the
experiment. Both experiments were approved by the local ethics committee.

7.3.2   Setup

Both experiments were performed using the Omega.3, as shown in Fig. 7.2 (ForceD-
imension, Switzerland). This is an impedance-controlled haptic device, with an
end effector with three translational degrees of freedom. In this experiment, the
movement of the device was restricted to one dimension, which was in the horizontal
plane, parallel with the frontal axis of the participant. Thus, participants could only
make left-right movements along a line of 20 cm. The middle of the device was
aligned with the body midline of the participant. The gravity of the device was
compensated using the Force Dimension DHD anti-gravity compensation scheme,
with a mass parameter of 0.06 kg. To provide higher transparency, a closed-loop
force control law was implemented in the device by adding a force sensor at the
end effector. This customization was performed previous to our experiment, and is
described in detail in Gurari and Baud-Bovy [9]. The control loop was implemented
as follows:

$$f_{cmd} = f_d - k_f (f_m - f_d). \tag{7.21}$$

In this control loop, f_{cmd} is the force command sent to the motors, f_m is the force
measured by the force/torque sensor, f_d is the desired force based on the desired

elastic force field, and k_f is the feedback gain, which was 10 in our experiment. The measured force (f_m) was passed through an exponential filter with a time constant of 0.004 s. The force command (f_{cmd}) was updated with a frequency of 1 kHz. Another customized feature was the addition of two pressure sensors to the handle, to measure the participant's grip force. For more information about the customization of the device, see Gurari and Baud-Bovy [9].

The participants always used their dominant hand to hold the handle of the haptic device, while they pressed the button on a response box with their other hand. To ensure a comfortable arm position, the chair height was adjusted to position the lower arm in the horizontal plane. Above the haptic device, at eye height, a small screen (Lilliput UK, 7″) provided visual information about grip force in all experiments. In Experiment 1, the screen was also used to provide visual feedback about hand position in one of the conditions. Vision of the hand was removed by placing a horizontal occluder below the feedback screen.

7.3.3 Protocol

In both experiments, the task for the participants was to find the position in the force field where the force was 0. Each trial started with a homing phase, in which the device guided the participant to the start position of that trial (see below for a description of the pseudo-random selection of start positions). After the homing phase, the force field was applied gradually. When the force field was at complete strength, a beep indicated that the participants could start exploring the force field. When the participants felt they had reached the desired position, they pressed the button of the response box to confirm the position. Then the experimenter started a new trial. There was no time limit and participants were told that there was no need to respond as quickly as possible. To make sure that the participants kept a firm but gentle grip on the handle throughout the experiment, with a comparable grip force across participants, visual feedback of the grip force was provided through a dot on a screen. The color of the dot corresponded to the grip force: when grip force was within the range of 0.25–1.5 N, the dot was green, while it turned red when the grip force was too high, and blue when it was too low. Participants were asked to make sure that the dot stayed green throughout each trial.

In Experiment 1, the elastic force field consisted of 2 linear springs, oriented right and left of the central position. The force field was oriented outwards, so a position farther from the central position corresponded to a stronger force away from the central position. This was done to avoid a 'letting go' strategy, in which participants would become so compliant that the machine would take them to the central position automatically. Because all the forces were fairly weak and the grip force was constantly monitored, this was not very likely, but we still wanted to avoid this situation. The task for the participants was to find the central position, which was described in the instructions as 'the position where the force became 0' or 'the position where the force changed direction'. Both descriptions were given to the participants. In half of the trials, visual feedback of the hand position was provided, using a moving dot on a horizontal white line (14 cm), which corresponded to the complete workspace (20 cm). On the horizontal line, 10 equidistant vertical lines were placed to provide some visual anchors. In the trials with visual feedback, the position of the dot on the line, which was also the grip force feedback dot, was an accurate representation of the actual hand position. In the other half of the trials, the dot was placed at a random position along the line and was stationary throughout the trial. It is important to note that in none of the trials, the visual feedback provided any information about the position of the force field. In most conditions, the force field was asymmetric, so the pairs of springs had different stiffnesses. The asymmetric spring stiffness pairs used in the experiment (left spring&right spring) were: 16&4, 16&8, 8&4, 4&8, 8&16, 4&16 N/m. In one condition, both spring stiffnesses were 8 N/m, which was the only symmetric condition.

For each combination of spring stiffnesses, the central position of the force field was varied by using a random position in one of the three regions, which were defined with respect to the middle of the workspace: from −3 to −1 cm, from −1 to 1 cm and from 1 to 3 cm. The central position of the force field was chosen at a random position in each of the three regions. The start position of the device was chosen at a random position in a region between 1 and 3 cm from the central position of the force field. Each random position was used for 2 start positions: one at the random distance to the left of the force field and one at the same distance to the right of the force field. So, this resulted in 3 regions × 2 start sides = 6 trials per condition. We used 7 combinations of spring constants (6 asymmetric pairs and one symmetric force field) and 2 feedback conditions (feedback present and absent), resulting in 84 trials per participant. These trials were divided into 3 blocks, resulting in a total measurement time of maximally one hour. In between the blocks, the participants could rest their arm to avoid fatigue.

In Experiment 2, the force field was uni-lateral, so only one linear spring was present. In this experiment, the force field was oriented inwards, so a position farther from the central position corresponded to a stronger force towards the central position. We could no longer use the instruction to 'find the position where the force changed direction'. Instead, we presented the task by telling participants that there was a virtual object of which they had to find the edge. Since visualizing an object is easier when it pushes you out instead of dragging you in, we chose to invert the direction of the force field. By setting a minimum grip force level, participants could not 'let go' to find the center of the force field in Experiment 2 either. The instructions for Experiment 2 were to 'to find the edge of the object, which pushes you out as soon as you enter it' or 'to find the position where the force just turns 0'. The spring constants were again 4, 8 or 16 N/m. The only visual feedback present in all the trials was the grip force dot, placed at a random horizontal position on the screen, while the vertical position was the same as in Experiment 1.

Again, we varied the position of the center of the force field by choosing random positions within the 3 defined regions. The start position of the device was chosen at a random position in a region between 1 and 3 cm from the central position of the force field and was always positioned outside of the object. We used 2 repetitions, resulting in 3 regions × 2 repetitions = 6 trials per condition. Since there were 6 conditions, the total number of trials was 36 per participant. The total experiment took about 20 min. Halfway into the experiment, the participants were asked if they wanted to have a break. Some participants did take a break, while others completed the experiment in a single block. During the experiment, force and position data were recorded with a frequency of 250 Hz. Prior to both experiments, some practice trials were performed to familiarize the participants with the task.

For both experiments, the stiffness model and the bisection model give different predictions. These predictions are illustrated in Fig. 7.1. For Experiment 1, a symmetric force field would yield an end position at the central position (H1) for both hypotheses. When an asymmetric force field is used, the stiffness model would still predict no bias (H1), while the bisection model predicts a bias towards the weaker force field (H2). For Experiment 2, if the stiffness model were used, an end position at the central position would be predicted (H1). However, when the bisection model were used, which is actually a one-sided model in this case, it predicts that the participant perceives the edge of the force field to be positioned at the force threshold (H2). So, the bisection model predicts a bias towards the spring, which would result in a negative bias for a spring on the left side and a positive bias for a spring at the right side.

7.3.4 Device Performance

For both experiments, some general characteristics describing the performance of the device were calculated first. For a general impression of the type of movements participants made, see Fig. 7.3. An important measure of the performance of the device is the difference between the desired and the measured force. Since force feedback is never completely transparent, these measures are never completely the same. In Experiment 1, the median force error across participants was 0.0013 N. In Experiment 2, the median force error was 0.0018 N, so the errors were very well centered around 0. The median RMS of the force error was 0.021 N for both experiments, so the errors were also fairly small. The calculation of the RMS was only based on data acquired within the linear part of the force field, which was between −6 and 6 cm with respect to the centre of the workspace. Another important measure is the position of the device at the zero-crossing of the force, as the position where the force changes direction is somewhat different in each movement. Moreover, because of friction and mechanical side-effects, there is a slight difference in force feedback between moving to the right and moving to the left. As the task for the participants was to find the zero-position of the force, this measure is very relevant. In Experiment 1, the median standard deviation of the zero-crossings, averaged per trial and then across trials, was 0.32 cm. For

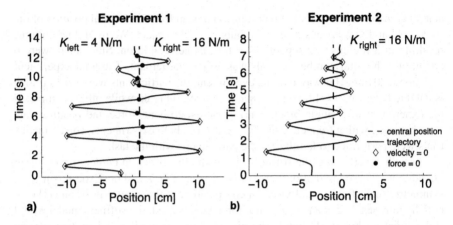

Fig. 7.3 Typical examples of movement trajectories in both experiments. (**a**) Typical example of Experiment 1. The diamonds indicate a change in movement direction, at which the velocity is 0. The black dots indicate a change in force direction, at which the force is 0. The dashed line indicates the centre of the force field, so the positions where the force is 0 are very close to this line. In this trial, the participant estimated the centre of the force field to be left of the true centre, since the trajectory ends left of the dashed line. This means that the bias is towards the weaker spring. (**b**) Typical example of Experiment 2. The dots are omitted in this graph, because the zero-force crossings are not well defined, since the force is 0 everywhere left of the central position. In this trial, there was only a spring on the right side of the workspace (which is the positive direction). The end position shows a bias towards this direction, so the end position lies slightly inside the object

Experiment 2, the zero-crossings were less well-defined, since the desired force was zero everywhere outside of the object. So, for Experiment 2, the zero-crossings were not calculated. The small standard deviation of the zero crossings shows that the desired and calculated central position of the force field matched closely. These measures together show that the performance of the device was good enough to be able to differentiate between the two different hypotheses, since biases were predicted to be in the range of about 0.5–3 cm.

7.3.5 Data Analysis

To first investigate the data qualitatively, some general characteristics to describe the movement patterns of the participants were calculated: movement time, total movement distance, number of times the movement direction was changed, and movement distance between these changes. All measures were first calculated for each trial and then averaged across trials and participants. As there were large differences between participants in their movement strategy, the variability in these measures was very large and did not look like a Gaussian distribution. Therefore, the median of the data (and not the mean) will be presented to describe the results of the general characteristics across participants.

The results for the end positions were more consistent across participants, so the results based on these data could be represented using mean ± standard error across participants. From the data sets of each participant, the bias and standard deviation were calculated for each condition. To calculate the bias, we used the difference between the central position and the end position of the movement, in which the latter was defined as the average position of the last 100 ms of the trial. The standard deviation was calculated across trials per condition and participant, so this represented the variability in the answers of each participant for each condition. For Experiment 1, for each trial, the positions at which the measured force changed direction were calculated, which we called zero-crossings. The mean position of these zero-crossing was used as the central position of the force field for that trial in further analyses, in order to use the best possible estimation of the force field that participants experienced, although the difference between the pre-defined and measured central position was very small. Since the zero-crossings were less well-defined for Experiment 2, they were not used to determine the central position in this experiment, but the pre-defined central positions were used for further analyses.

An outlier analysis was performed on the basis of 2 criteria. Firstly, when participants used too much grip force or moved very fast, the setup sometimes produced high-frequency noise. When this happened for more than 0.5 s consecutively, the trial was rejected, which was the case for 6 trials (0.60%) in Experiment 1 and 1 trial (0.23%) in Experiment 2. The second criterion was based on the consistency of the data. For each condition, trials were rejected that were more than 5 standard deviations away from the mean of the condition, when the mean was calculated without that particular trial. On the basis of this criterion, 10 trials (0.99%) were rejected in Experiment 1 and 5 trials (1.2%) were rejected in Experiment 2.

To assess the correspondence between predictions from the bisection model and measured biases, the biases were first predicted based on detection thresholds from literature (0.05–0.1 N). The measured biases were averaged across visual feedback conditions for easy comparison to the predicted data. If the bisection model holds, the data should fall within the predicted range. If the measured biases were lower than the predicted ones, this would be in favor of the stiffness model. All further analyses and the fitting procedures were performed on data separated according to visual feedback condition.

To assess the effect of the different spring stiffnesses and of the visual feedback, repeated measures ANOVAs were performed, for the biases and the standard deviations separately. When the sphericity criterion was not met, Greenhouse-Geisser correction was used. The biases and standard deviations were averaged across participants before fitting a model. The goodness-of-fit of the model was assessed using an R^2 value.

7.4   Results

7.4.1   Measured and Expected Biases

To assess the validity of the models, the biases were first predicted based on detection thresholds from literature (0.05–0.1 N, as reported in Baud-Bovy and Gatti [3]). These predicted values are shown as gray bars in Fig. 7.4, as well as the measured data. For Experiment 1, the measured values mostly fall in the predicted range. The measured values in Experiment 2 are even a bit higher than expected. These observations are not in accordance with the stiffness model, so we did not use this model in the rest of the analysis. In the next subsections, the measured data and the fits made using the bisection model are discussed in more detail.

7.4.2   Experiment 1

The performance of the participants was first characterized by investigating the movement trajectories qualitatively. Generally speaking, participants first made large sinusoidal movements, after which they gradually decreased their movement amplitude to 'zoom in' to the position where the centre of the force field was located. A typical example of such a trajectory is given in Fig. 7.3a. Across all trials and participants, the median movement time was 10.9 s, while the total moved distance

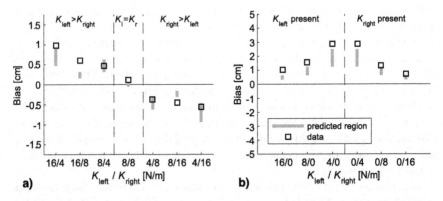

Fig. 7.4 Measured data and biases predicted using the bisection model, based on force detection thresholds from literature. The squares represent the mean of the measured biases (averaged across visual feedback conditions), while the gray lines represent the predictions from the bisection model for the range of force detection thresholds from literature (0.05–0.1 N). (a) Measured data and predicted biases for Experiment 1. (b) Measured data and predicted biases for Experiment 2. If the measured biases were lower than the predicted ones, this would indicate that the stiffness model is the most likely explanation of the data. For both experiments, this is clearly not the case, so the bisection model is better at describing the data

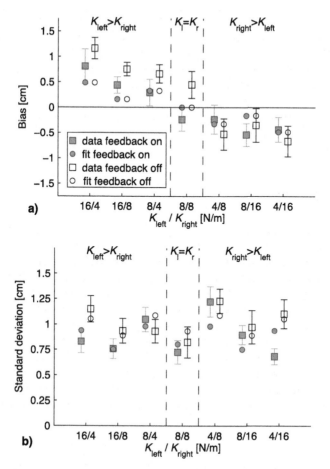

Fig. 7.5 Data and fits of Experiment 1. The squares with error bars represent the mean±one standard error, while the dots represent the fit of the bisection model to the data. On the horizontal axis, the stiffness pairs of the asymmetric force field are indicated. The filled symbols indicate the condition in which visual feedback is present, while the open symbols show the condition without visual feedback. (**a**) Biases of Experiment 1. All biases are oriented towards the weaker force field. Note the effect of combination of spring stiffnesses on the size of the bias. There was no significant effect of visual feedback type on the bias. (**b**) Standard deviations of Experiment 1. Note that the standard deviation is significantly higher for the condition without visual feedback, compared to the condition with visual feedback

was 50.6 cm. During each trial, participants changed movement direction 11.1 times, while the median distance between two changes in direction was 5.9 cm.

The analysis of the endpoints showed biases towards the weak side of the force field, as can be seen in Fig. 7.5a. So, when the spring on the right (left) side was weaker, the biases were positive (negative) and thus rightwards (leftwards). Moreover, when the difference between the two springs was larger, the bias was also larger. The repeated measures ANOVA on the biases showed a significant main effect of force field stiffness ($F_{1.9,66} = 12$, $p \leq 0.001$), but not of visual feedback type ($F_{1,11} = 1.5$, $p = 0.25$). For the standard deviations, which are shown in Fig. 7.5b, the repeated measures ANOVA also showed a significant main effect of force field stiffness and a significant main effect of visual feedback type ($F_{6,66} = 3.5$, $p = 0.0043$ and $F_{1,11} = 5.0$, $p = 0.047$, respectively). So, even though there was no difference in bias between the two types of visual feedback, the standard deviation across trials within participants was smaller when the visual feedback dot was moving.

7.4.3 Experiment 2

In this experiment, participants generally responded faster and made more asymmetric movements than in Experiment 1, as can be seen from the typical example in Fig. 7.3b. Their median movement time was 8.5 s, while the total moved distance was 39.2 cm. During each trial, participants changed movement direction 8.2 times, while the median distance between two changes in direction was 4.5 cm.

The analysis of the endpoints showed significant biases, as can be seen in Fig. 7.6a. All mean biases were positive (negative), which means rightwards (leftwards), for a force field on the right (left). So, the participants perceived the center of the force field to be slightly inside the object. The repeated measures ANOVAs showed a significant main effect of type of force field on both the biases and standard deviations ($F_{5,55} = 23.7$, $p < 0.001$ and $F_{5,55} = 9.93$, $p < 0.001$, respectively). This is also reflected in Fig. 7.6: both the biases and the standard deviations are larger for smaller force field stiffnesses.

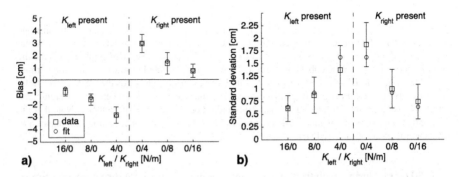

Fig. 7.6 Data and fits of Experiment 2. The squares with error bars represent the mean±one standard error, while the dots represent the fit of the bisection model to the data. On the horizontal axis, the stiffness of the uni-lateral force field is indicated (a 0 is noted for the side on which there was no spring). (**a**) Biases of Experiment 2. All biases are oriented towards the spring. Note that the bias increases for weaker springs. (**b**) Standard deviations of Experiment 2. The standard deviations are also larger for weaker springs

Table 7.1 Fitted values
using the bisection model, for
Experiments 1 and 2

Experiment	F_{th} [N]	σ_F [N]	$\sigma_{P_{fa}}$ [cm]	$\sigma_{P_{fp}}$ [cm]
1	0.085	0.052	1.2	0.93
2	0.12	0.062	0.51	(–)

7.4.4   Model Fits

The data of both experiments showed strong biases that were highly dependent on
the stiffness of the force fields. Figure 7.4 already illustrated that the measured data
are not in accordance with the stiffness model, so only the fits of the bisection model
are discussed in this subsection. For Experiment 1, the mean biases and standard
deviations were fitted together for both feedback conditions (See Eq. 7.10), while
using the fitting parameters force threshold (F_{th}), noise on the force threshold (σ_F),
noise on the position in the condition in which visual feedback was present ($\sigma_{P_{fp}}$),
and noise on the position in the condition in which visual feedback was absent ($\sigma_{P_{fa}}$).
This yielded a value of 0.085 N for the force threshold and 0.052 N for the noise on
the force threshold. This force threshold falls exactly in the range of expected force
thresholds (between 0.05 and 0.1 N), so this is not in accordance with the stiffness
model. The noise on the position data was lower for the condition in which visual
feedback was present than for the condition in which it was absent. For the former
condition, the fitted positional noise was 0.93 cm, while it was 1.2 cm for the latter.
The goodness-of-fit was good, with an R^2 of 0.87.

For Experiment 2, the mean bias and standard deviations were also fitted together
(see Eq. 7.10), while using the fitting parameters force threshold (F_{th}), noise on the
force threshold (σ_F), noise on the position (σ_P). This yielded values of 0.12 N for
the force threshold and 0.062 N for the noise of the force threshold. The fitted force
threshold is close to the range described in literature, but it does not exactly match.
However, if the stiffness model were true, thresholds beneath 0.05 N are expected,
so the threshold being higher than the expected value is not in accordance with the
stiffness model. The fitted noise on the position data was 0.51 cm. The goodness-
of-fit was excellent, with an R^2 of 0.99. An overview of all the fit values is given in
Table 7.1.

7.5   Discussion

In both experiments, we found significant stiffness-dependent biases, that were
between (Experiment 1) or at (Experiment 2) the force detection threshold expected
from literature, which is in accordance with the hypothesis that participants base
their estimation of the centre of a force field on the position(s) where the force
reaches the threshold level. If the stiffness model were used, much smaller biases
would have been found. In Experiment 1, the biases were all oriented towards the

weaker side of the force field. The values of the combination of spring stiffnesses significantly influenced both the biases and the standard deviation of the biases. This influence was also present in the model fits, as can be seen from the model fits in Fig. 7.5. The manipulation of visual feedback only significantly influenced the standard deviation and not the biases themselves, although the effect on the standard deviations was small. The bisection model also captured this feature, as it also produced a lower fit value for the positional noise in the condition in which visual feedback was present.

In Experiment 2, the biases were all oriented towards the force field, so all biases were positioned inside the virtual object. The stiffness of the force field significantly influenced both the biases and the standard deviations: a smaller stiffness caused larger biases and standard deviations. All these observations are in line with the predictions from the bisection model, while they do not support the prediction of biases smaller than the detection threshold from the stiffness model.

Together, these experiments not only provide strong evidence against the use of stiffness information in this task, but they also provide evidence that the bisection model is able to correctly predict biases and standard deviations for this task. So, it seems that participants indeed only use the position(s) where the force reaches the threshold to solve the task. The fitted values are comparable to values found in previous experiments using this model [2, 4]. Moreover, the fitted force thresholds are very close to thresholds measured in other behavioral experiments [3], even though our fitted thresholds are only based on the observed biases and standard deviations. It is puzzling that the positional noise in Experiment 2 is lower than that of both conditions of Experiment 1. An explanation could be that the integration between a left and a right position, which needs to be done in Experiment 1, adds some noise to the position estimate, which is currently not incorporated in the model. The force threshold is also a bit higher in Experiment 2. This could be caused by participants using a cautious strategy: it is safer to move a bit more inside the object when you want to be certain that you have localized it, which would result in an increase in the fitted force threshold. In Experiment 1, this strategy does not work, because there is a force field on both sides. However, the difference between the force thresholds is not very large.

To be able to mathematically describe the models, some assumptions needed to be made. One of them is the assumption of using an offset in Stevens' law (see Eq. 7.2, the offset is the x_0 term). We assumed this offset to be equal to the position of the force detection threshold. Decreasing this offset would have decreased the biases predicted from the stiffness model, while it would not have affected the predictions from the bisection model. Decreasing this term to 0 would have led to a prediction of no biases at all for the stiffness model, so the stiffness model would still have been rejected when decreasing the offset. Another assumption is that β is 0.8. If we had chosen a much higher β, the biases predicted from the stiffness model would have been much larger than the ones expected from the force detection thresholds from literature. So, for β smaller and larger than 1, we can reject the stiffness model. However, when assuming both $\beta = 1$ and $x_0 =$ the position of the detection threshold, no distinction between the models can be made. To assess the validity of the models under these assumptions, future research using other paradigms that would be able to discriminate between the models under these assumptions would be needed, like force fields composed of non-linear springs.

Even though our results are in accordance with previous experiments testing the idea that participants base their estimate of the centre of a force field on the position(s) where the force reaches the threshold level, they are still slightly puzzling from the perspective of the sources of information that participants have. The results clearly show that participants do not use an estimate of the stiffness, even though they could have had access to this information if they wanted to, since they were exploring the force field anyway and thus were experiencing changes in position and the accompanying changes in force. By not using the stiffness information, they made large errors, which is usually something you want to avoid. This implies that force and position are not integrated in this task, even though previous research has shown that humans are able to do this [e.g. 6].

One explanation of this phenomenon could be that, to be able to use stiffness information, participants would have to make certain assumptions. For instance, they would need to assume that the spring is linear in the range where the forces are imperceptible. In nature, objects hardly ever behave like a linear spring, so this might explain why this assumption is not a logical one to make. However, participants should be able to assess that the part of the force field with forces above threshold level is linear, so then the assumption of a completely linear spring would not be very unlikely.

Another explanation is that stiffness information might not be a very reliable cue, since humans show poor performance in stiffness discrimination tasks [13]. Tan et al. [22] even argue that humans do not use stiffness information at all. They showed that, in their task of squeezing two plates in a pinch grasp, terminal force cues and work cues were the primary source of information that participants used to estimate the stiffness.

A third explanation could be that there is a cost to acquiring stiffness information, which is larger than the cost of using the strategy described in the bisection model. Since participants never received feedback about their performance, they did not know that they were making errors. If the force field would have been symmetric, the bisection model would have predicted no errors, so in that case, it could have been a smart strategy to choose. Acquiring stiffness information could be costly, because the process is very serial, so information needs to be acquired and compared over time. When only using the position where the force reaches the threshold, the task of finding that position might be of a serial nature, but the information that needs to be stored is only one or two position(s). In several studies, it has been suggested that serial strategies are more costly than parallel ones [5, 17]. There are also examples of situations in which haptic information is neglected when it is added to a task. For example, Heuer and Rapp [10] describe that added haptic information is neglected in a visuo-motor rotation task. In a next study, they describe a deterioration in learning of a visuo-motor rotation when augmented haptic feedback is added, so the neglect of haptic information might be a functional strategy in this task to avoid deterioration in learning [11]. The same process might be happening in our experiment: even though stiffness information is provided, participants neglect it and choose the 'easy option' described by the bisection model to solve the task. This suggests that the brain might be able to choose if it integrates information or not, based on the possible costs and benefits of the integration. However, irrespective of which explanation is the correct one, the bisection model seems to adequately describe the outcome of the decision process of the participants.

The knowledge acquired in this experiment also has practical applications, such as for the design of haptic guidance in tele-operation applications [1]. In this technique, force fields are often used to guide the operator towards a target, which is a position at the minimum of the force field. Experiment 1 shows that it could be beneficial to add visual feedback about hand position, even if the feedback does not provide task-relevant information. Experiment 2 suggests that it might not be smart to position the target at the minimum of the force field, since operators actually hardly overshoot the target that far that they enter the part of the force field that is above threshold level behind the target. So, in practice, force fields in haptic guidance resemble a uni-lateral situation, in which we found large biases. Therefore, it might be better to ensure that the force is at threshold level when operators reach the target. Obviously, follow-up experiments would be needed to test if this suggestion indeed increases operator performance.

7.6   Conclusion

When participants are asked to estimate the centre of a force field, they show a stiffness-dependent bias towards the weaker spring in an asymmetric force field. In a uni-lateral force field, they show a stiffness-dependent bias towards the direction of the spring. When providing visual feedback about hand position, this decreases the variability in the responses, even if it does not provide any task-relevant information. This study provides evidence against the hypothesis that participants use stiffness information to find the centre of a force field. The results are in agreement with the idea that participants base their estimation of the centre of a force field on the position(s) where the force reaches the threshold level, which is described mathematically in the bisection model. This shows that force and position information are not integrated in this task.

References

1. Abbink DA, Mulder M, Boer ER (2012) Haptic shared control: smoothly shifting control authority? Cogn Tech Work 14(1):19–28
2. Baud-Bovy G (2014) The perception of the centre of elastic force fields: a model of integration of the force and position signals, chap 7. Springer series on touch and haptic systems. Springer, London, pp 127–146

3. Baud-Bovy G, Gatti E (2010) Hand-held object force direction identification thresholds at rest and during movement. In: Kappers AML, Van Erp JBF, Bergmann Tiest WM, Van der Helm FCT (eds) Haptics: generating and perceiving tangible sensations. Lecture notes in computer science, vol 6192. Springer, Berlin/Heidelberg, pp 231–236

4. Bocca F, Baud-Bovy G (2009) A model of perception of the central point of elastic force fields. In: Proceedings of the 3rd joint world haptics conference and symposium on haptic interfaces for virtual environment and teleoperator systems (WHC), Salt Lake City, pp 576–581

5. Dopjans L, Bülthoff HH, Wallraven C (2012) Serial exploration of faces: comparing vision and touch. J Vis 12(1:6):1–14

6. Drewing K, Ernst MO (2006) Integration of force and position cues for shape perception through active touch. Brain Res 1078(1):92–100

7. Ernst MO, Banks MS (2002) Humans integrate visual and haptic information in a statistically optimal fashion. Nature 415(6870):429–433

8. Ernst MO, Bülthoff HH (2004) Merging the senses into a robust percept. Trends Cogn Sci 8(4):162–169

9. Gurari N, Baud-Bovy G (2014) Customization, control, and characterization of a commercial haptic device for high-fidelity rendering of weak forces. J Neurosci Methods 235:169–180

10. Heuer H, Rapp K (2012) Adaptation to novel visuo-motor transformations: further evidence of functional haptic neglect. Exp Brain Res 218(1):129–140

11. Heuer H, Rapp K (2014) Haptic guidance interferes with learning to make movements at an angle to stimulus direction. Exp Brain Res 232(2):675–684

12. Jacobs RA (1999) Optimal integration of texture and motion cues to depth. Vis Res 39(21):3621–3629

13. Jones LA, Hunter IW (1990) A perceptual analysis of stiffness. Exp Brain Res 79(1):150–156

14. Korman M, Teodorescu K, Cohen A, Reiner M, Gopher D (2012) Effects of order and sensory modality in stiffness perception. Presence Teleop Vir Environ 21(3):295–304

15. Kuschel M, Di Luca M, Buss M, Klatzky R (2010) Combination and integration in the perception of visual-haptic compliance information. IEEE Trans Haptic 3(4):234–244

16. Landy MS, Maloney LT, Johnston EB, Young M (1995) Measurement and modeling of depth cue combination: in defense of weak fusion. Vis Res 35(3):389–412

17. Loomis JM, Klatzky RL, Lederman SJ (1991) Similarity of tactual and visual picture recognition with limited field of view. Perception 20(2):167–177

18. Robles-De-La-Torre G, Hayward V (2001) Force can overcome object geometry in the perception of shape through active touch. Nature 412(6845):445–448

19. Srinivasan MA, LaMotte RH (1995) Tactual discrimination of softness. J Neurophysiol 73(1):88–101

20. Stevens S (1957) On the psychophysical law. Psychol Rev 64(3):153–181

21. Stevens JC, Marks LE (1999) Stevens's power law in vision: exponents, intercepts, and thresholds. In: Proceedings of the fifteenth annual meeting of the international society for psychophysics (Fechner Day 1999), pp 87–92

22. Tan HZ, Durlach NI, Beauregard GL, Srinivasan MA (1995) Manual discrimination of compliance using active pinch grasp: the roles of force and work cues. Percept Psychophys 57(4):495–510

23. Van Beers RJ, Sittig AC, Denier van der Gon JJ (1999) Integration of proprioceptive and visual position-information: an experimentally supported model. J Neurophysiol 81(3):1355–1364

24. Van Beek FE, Bergmann Tiest WM, Mugge W, Kappers AML (2015) Haptic perception of force magnitude and its relation to postural arm dynamics in 3D. Sci Rep 5:18,004

Chapter 8
Visuo-Haptic Biases in Haptic Guidance

Abstract Visuo-haptic biases are observed when bringing your unseen hand to a visual target. The biases are different between, but consistent within participants. We investigated the usefulness of adjusting haptic guidance to these user-specific biases in aligning haptic and visual perception. By adjusting haptic guidance according to the biases, we aimed to reduce the conflict between the modalities. We first measured the biases using an adaptive procedure. Next, we measured performance in a pointing task using three conditions: (1) visual images that were adjusted to user-specific biases, without haptic guidance, (2) veridical visual images combined with haptic guidance, and (3) shifted visual images combined with haptic guidance. Adding haptic guidance increased precision. Combining haptic guidance with user-specific visual information yielded the highest accuracy and the lowest level of conflict with the guidance at the end point. These results show the potential of correcting for user-specific perceptual biases when designing haptic guidance.

Previously published as:
F.E. van Beek, I.A. Kuling, E. Brenner, W.M. Bergmann Tiest & A.M.L. Kappers (2016)
Correcting for visuo-haptic biases in 3D haptic guidance
PLoS ONE 11(7): e0158709

8.1 Introduction

In remote, dangerous or space-limited environments, teleoperation systems show great potential. In these systems, the operator uses a master device to control a slave system, which is placed in the environment. The master and the slave can be separated over a large distance. A consequence of the separation between operator and environment is the lack of natural feedback to the operator, so these systems are usually equipped with devices to provide haptic feedback in such a way that it resembles natural haptic feedback [7]. In addition to recreating the natural scene, these feedback devices could also be used to augment reality, of which the concept of haptic guidance is an example. In haptic guidance, forces inform the operator

© Springer International Publishing AG 2017
F.E. van Beek, *Making Sense of Haptics*, Springer Series on Touch and Haptic
Systems, https://doi.org/10.1007/978-3-319-69920-2_8

about the task objective by guiding the operator towards the target location [1]. This concept can, for instance, be used in a peg-in-hole task, in which a peg needs to be placed at a certain position and orientation to allow it to enter the hole. To guide the user to the correct position, haptic guidance in the form of an attractive force field, which has zero force at the desired position, can be used. One of the big challenges is to design these guidance forces in such a way that they are easy in use. In order to do this, knowledge on human perception is essential. One of the potential problems could be that human perception is not veridical, as is evident from various biases within and between modalities. Perceptual biases can be consistent across participants, such as the radial-tangential illusion [5], or they can be consistent within, but not between participants, such as the user-dependent error-patterns in the perception of force direction [21]. In a natural situation, such as trying to grasp an object, humans automatically correct for such biases, for instance by using both visual and haptic information. In teleoperation situations, this automatic correction might not be possible, for instance because operators do not see their hands while performing actions and thus cannot align their visual and haptic information. For haptic devices, it has been shown that correcting the mapping between operator and slave movements, by using parameters that are consistent across participants, increases user performance [15]. The question in our study was if it is also useful to correct for user-specific biases in the design of haptic guidance, by making user-specific adjustments.

To test this question, we used the well-known paradigm of visuo-haptic biases [18]. When participants are asked to bring their unseen hand to a visual target position, they make large errors, which are called visuo-haptic biases. These errors are idiosyncratic: they are consistent within, but not between participants [8, 16, 17]. Even when these biases are measured a month later, they are still consistent [12] and they are not affected by imposed forces or torques [9, 10]. When the targets are restricted to a horizontal plane, the biases are larger when the arm is further away from the body along the body midline [4, 23] or when the targets are positioned further to the left or right [6]. These natural mismatches are far less likely to be corrected for by the nervous system than are externally imposed mismatches [24].

We used the existence of visuo-haptic biases to test for the effectiveness of user-specific adjustments of haptic guidance, in a task in which participants were asked to localize visually presented targets without seeing their hand. When perception is biased, there is a difference between the physical and the perceived properties of a stimulus. In the case of visuo-haptic biases, there is a difference between the visual perception of the location of the stimulus and the haptic perception of that location. The practical implication of this is that if haptic guidance is directed towards a visual target, participants do not perceive it to be directed to that location. If the biases are known, it is possible to correct for them by adjusting the feedback information. In this way, the visual and haptic information could be made perceptually consistent, rather than physically consistent.

In a previous experiment, it was found that correcting for visuo-haptic biases in a 2D situation improved operator performance [11]. In the current study, we used

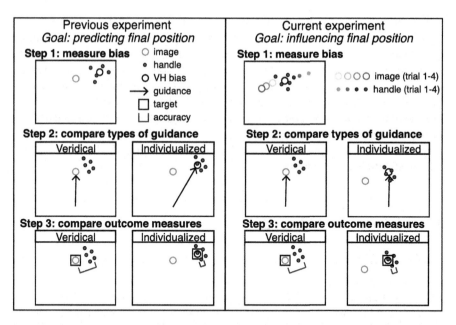

Fig. 8.1 Illustration of the protocol of the previous and current experiment. In the previous experiment (left box), the goal was predicting where participants would position the handle, when being asked to move to a postion. In order to do that, the direction of the guidance force was adjusted, while the position of the visual image was constant throughout the experiment. The outcome measure was how closely the mean end position of the handle matched the center of the guidance force field. In the current experiment (right box), the goal was influencing where participants would position the handle. To achieve that, the visual image position was shifted to match the visuo-haptic bias, while the guidance force was always directed towards the original visual position. The outcome measure was how closely the mean handle position matched the original visual image position (and thus the center of the guidance force field)

the same task of pointing to visual targets with the unseen hand. There are two important differences between the previous study and the current one, which are illustrated in Fig. 8.1. These differences are: (1) in the previous study, the concept was investigated in 2D, whereas we used a 3D setup in the current study; and (2) in the previous study, the information was corrected by changing the position of the haptic guidance, while in the current study, we adjusted the position of the visual information. So, in the previous experiment, the visual image was always presented at the same position, and veridical haptic guidance, towards the target position, was compared to haptic guidance towards the target position shifted by the measured bias. In the current study, we shifted the visual image, while the haptic guidance and the haptic target position were constant. In both methods (shifting the guidance position or shifting the position of the visual information), the aim was to reduce the amount of conflict between the modalities. The approach was somewhat different: in the previous experiment, participants reached the target position shifted by the visuo-haptic bias and not the original target position when performing consistently. When shifting the visual image, which was done in the current experiment, participants do reach the target position when performing consistently.

In summary, the aim of this study was to test the effectiveness of user-specific adjustments in haptic guidance, for which we used the paradigm of visuo-haptic biases in a 3D situation. We first measured the visuo-haptic bias and then compared three conditions: (1) a condition with visual images that were adjusted to the user-specific bias, without haptic guidance, (2) a condition with veridical images, combined with haptic guidance, and (3) a condition with shifted visual images, combined with haptic guidance.

8.2 Materials and Methods

8.2.1 Participants

Twelve participants took part in the experiment. The calibration failed for one participant, which we only discovered upon completion of the experiment. Therefore, the data of that participant could not be used. The data of the other eleven participants were used in the analysis. Three participants were males and eight were females, nine were right-handed and two were left-handed. Their age was 30 ± 3 years (mean $\pm$ standard deviation). None of the participants had a history of neurological disorders, all had normal or corrected-to-normal visual acuity, and none were stereo blind. Prior to the experiment, they received written instructions and signed an informed-consent form. The experiment was approved by the Ethics Committee of the former Faculty of Human Movement Sciences (ECB).

8.2.2 Apparatus

The setup was a combination of a visual display to provide visual information in 3D, a haptic device to provide guidance forces and a position tracking system, as shown in Fig. 8.2. The visual display consisted of two CRT monitors (1096×686 pixels, 47.3×30.0 cm, refresh rate 160 Hz), which projected images onto two mirrors in front of the participant. By providing a different image for each eye, a 3D image could be created. Sousa et al. [20] provide a more elaborate description of the visual setup. Participants were seated on a height-adjustable chair to make sure that the mirrors were positioned at eye-height. The room was completely dark except for the light from the screens.

An Optotrak system (type 3020, Northern Digital Inc.) was used to track the position of 4 Infra-Red Emitting Diodes (IREDs) with a frequency of 250 Hz. Three IREDs were attached to a bite-board, and one IRED was attached to the handle of

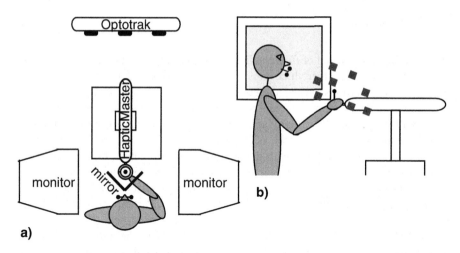

Fig. 8.2 Participant in the setup. The participant was seated between 2 monitors, which projected images onto the semi-silvered mirrors to present target cubes in 3D. The participant used the handle of the HapticMaster to point at the perceived target locations. In some conditions, the HapticMaster exerted guidance forces, while in other conditions, it only passively followed the movements of the participant. The position of the handle and the head orientation of the participant were tracked using the Optotrak system (cameras are shown as filled rounds on the Optotrak and markers are shown as grey dots). (**a**) Top view. (**b**) Side view with target locations depicted as gray cubes. For the precise coordinates of the target positions, see Table 8.1

the haptic device. During the experiment, participants wore the bite-board in their mouth. Prior to the experiment, we determined the position of the eyes relative to this bite-board, using a procedure described previously in Sousa et al. [19]. During the experiment, the markers on the bite-board were tracked constantly by the Optotrak in order to display correct images to the participants, even when they made head movements. In this way, the 3D information was realistic, while allowing participants to move their head during the experiment. The fourth IRED was used to track the position of the handle of the haptic device in the Optotrak coordinate frame. Since the orientation of the handle could not change, we could use a single marker to determine the position of the center of the handle.

The haptic device was a HapticMaster (Moog Inc.), a strong admittance-controlled haptic device, which can also record position and force data [25]. In this experiment, the device recorded force and position data with a sampling frequency of 256 Hz. Its inertial mass was set to 3 kg throughout the experiment, and it moved frictionlessly through its closed-loop control. In the conditions without haptic guidance, the device followed the movements of the participants passively, while only compensating for its own weight. In the other conditions, it exerted a guidance force on the hand of the participant. Participants held the ball-shaped handle by enclosing it with their dominant hands. Their non-dominant hand was used to press two buttons to proceed to the next trial. Prior to the experiment, the coordinate systems of the Optotrak and the HapticMaster were aligned.

8.2.3   Protocol

The experiment consisted of four conditions. All participants started with the same
condition, while the other three conditions were presented in counterbalanced
order. In all conditions, visual targets were presented using small green cubes
($1 \times 1 \times 1$ cm). The task for the participant was always the same: align the center
of the ball-shaped handle of the HapticMaster with the center of the visual target
cube. The position of the handle of the HapticMaster was never visible. When
the participants were satisfied with the alignment of the centers, they pressed two
spacebars with two fingers of their non-dominant hand to confirm the position
in both measurement systems. After pressing the spacebars, a new trial started
automatically from the end position of the previous trial. Participants were told that
in some conditions, the haptic device would exert forces to help them reach the
target. They were instructed that these forces would always help them to move in
approximately the right direction, and that they would not always be correct when
being close to the target. It was stressed that their task was to indicate the visible
target's position as accurately as possible, irrespective of the direction of the forces.
In all conditions, 8 target positions were used, which are shown in Fig. 8.2b and
in Table 8.1. Each target position was repeated 20 times per condition, resulting in
160 trials per condition. A pseudo-random order was used for target presentation:
the trials were divided into blocks, in which each target position was presented
once. Within these blocks, the order was randomized. The same target was never
presented on two consecutive trials. The 20 blocks of each condition were presented
consecutively, without breaks.

In the first condition (VA: constantly adjusting visual information only), the aim
was to find the visual image position ($\mathbf{x}_{v,k}$) of target k ($k = 1, 2, \ldots, 8$) at which the
position of the haptic representation ($\mathbf{x}_{h,k}$) of that image coincided with the intended
target position ($\mathbf{x}_{t,k}$). To achieve this, the position of the visual image was adjusted
on trial n ($n = 1, 2, \ldots, 20$) based on the measured visuo-haptic bias. On the
first trial, the visual target was presented at the target position ($\mathbf{x}_{t,k}$). Participants
placed the handle at their haptic representation of the visual image position ($\mathbf{x}_{h,k,n}$),
which was usually different from the actual target position. The error between the

Table 8.1 Target positions,
with respect to the middle of
the workspace

Target	x [cm]	y [cm]	z [cm]
1	−7.5	−1.5	−6.0
2	2.5	1.5	6.0
3	7.5	4.5	−2.0
4	−2.5	4.5	2.0
5	2.5	−4.5	−2.0
6	−7.5	1.5	2.0
7	7.5	−4.5	−6.0
8	−2.5	−1.5	6.0

haptically perceived position and the target position was calculated and the visual image position was adjusted in the next trial for each target position (k) separately, using:

$$\mathbf{x}_{v,k,n+1} = \mathbf{x}_{v,k,n} - 0.4(\mathbf{x}_{h,k,n} - \mathbf{x}_{t,k}). \tag{8.1}$$

In this way, the visual image $(\mathbf{x}_{v,k,n})$ was adjusted on each next trial $(\mathbf{x}_{v,k,n+1})$ in the direction opposite to the error between the end point of the participant's movement in the current trial $(\mathbf{x}_{h,k,n})$ and the target position $(\mathbf{x}_{t,k})$. As a result, the participant should be placing the handle at the target position $(\mathbf{x}_{t,k})$, after some trials. This allowed us to determine the visual position that corresponded perceptually with the intended haptic target position. The adaptive value of 0.4 was chosen, as this is an optimal parameter for some such adaptive tasks [22]. A typical example of data from this condition is presented in Fig. 8.3.

In the second condition (VS: shifted visual information only), the means of the visual image positions of the last 10 trials for each target position of condition VA were used as the visual image positions. The aim of this condition was to measure if these shifts of the visual image positions indeed resulted in participants indicating their end positions close to the target positions. Moreover, the perceptual precision could be determined in this way. A typical example of data from this condition is presented in Fig. 8.4 (left column).

In the third condition (VS+H: shifted visual information and haptic guidance towards the target location), haptic guidance forces were added to the situation in condition VS. These guidance forces were designed as an attractive force field, with

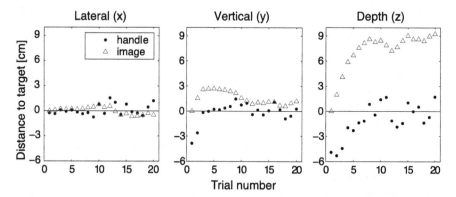

Fig. 8.3 Illustration of the adaptive procedure used in condition VA. The triangles show the positions of the visual image, which is initially presented at the target position, and the dots indicate the end positions of the device handle. In all directions, zero indicates the target position. The three panels show the projections of the data onto the cardinal axes for one target. The visual image is shifted so that the end positions of the handle will be closer to the target at the last trials. The mean position of the visual images in trials 11 through 20, which is fairly stable with some noise around the mean, was used as the position of the shifted visual image throughout the rest of the experiment

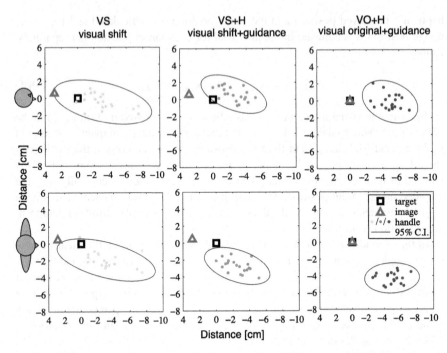

Fig. 8.4 Typical examples for conditions VS, VS+H, and VO+H for one target and one participant. The black square shows the target position. The gray triangle shows the position of the visual image. The dots indicate the end positions of the handle. The ellipse is a 2D representation of the 95% confidence ellipsoid. The first row shows a side view (i.e. horizontal axis shows depth direction, vertical axis shows vertical direction), and the second row shows a top view of the data (i.e. horizontal axis shows depth direction, vertical axis shows lateral direction). First column: condition VS. The visual image is shifted from the target position. The target square lies inside the confidence ellipsoid, but the ellipsoid is fairly large. Second column: condition VS+H. The visual image is again shifted, and haptic guidance towards the target position is added. The target square is still fairly close to the confidence ellipsoid. The volume of the ellipsoid is smaller than in VS. Third column: condition VO+H, in which the visual image is shown at the target location, while haptic guidance towards the target position is also added. Again, the volume of the confidence ellipsoid is fairly small, but now the ellipsoid is much farther from the target than in VS+H

a linear stiffness of 50 N/m, and with an upper limit of 3.5 N to avoid large forces. So, at the target position, the guidance force was 0, while the force towards the target increased linearly with the distance from the target. In this condition, the visual image and the position of the minimum of the guidance force field were not presented at the same physical position, but the shifted visual position was used to correct for the visuo-haptic bias. A typical example of data from this condition is presented in Fig. 8.4 (middle column).

In the fourth condition (VO+H: veridical visual information and haptic guidance towards the target location), haptic guidance forces were the same as in condition VS+H. The visual image was also presented at the target location, so all information was presented veridically. A typical example of this condition is presented in Fig. 8.4 (right column).

In between the conditions, participants took a break as long as they preferred, in which they were allowed to take the bite-board out of their mouth. Each condition took about 8 min, so the total experiment lasted about 45 min per participant.

8.2.4   Data Analysis

To analyze the data, 95% confidence ellipsoids were fitted to the end points of the participant's movements, for each target, condition and participant. To assess the accuracy of the participant's performance, the distances between the centers of the ellipsoids and the actual target positions were calculated. Hence, a larger number indicates a lower accuracy. These values were averaged across target positions prior to statistical analyses. A similar calculation was done for condition VA, but here the difference between the target position and the mean position of the images in the last 10 trials of condition VA was used, since this represents the visuo-haptic bias. So, this value indicates how far the visual image needed to be shifted in order for the participant to perceive the hand at the target location. The target locations, shown in Table 8.1, were only used as the visual image positions in condition VO+H and in the first trial of condition VA. In all other conditions, participants were asked to move towards a shifted visual image, while the accuracy measure was always based on how closely the mean end positions matched with the target locations.

To assess the precision of the participant's performance, the volumes of the ellipsoids were used, so a larger number indicates lower precision. Again, these values were averaged across target positions prior to performing statistical analyses. A repeated measures ANOVA was performed for both accuracy and precision data separately, on conditions VS, VS+H, and VO+H, to assess the effect of condition. Post-hoc tests with Bonferroni correction were used to compare all possible pairs of conditions. For all repeated measures ANOVAs, Greenhouse-Geisser correction was used when the sphericity-criterion was not met according to Mauchly's test.

The level of conflict between the haptic guidance and the participant's judgement was assessed by calculating the amount of guidance force that was still present when participants decided they had reached the target position. This measure is comparable to the accuracy measure, but the guidance force was limited to a maximum of 3.5 N, so, the distance measure was not limited, while the conflict measure had a limit of 3.5 N. It also provides an additional way of looking at the data, as this measure directly shows how much force was left on the handle when the participant decided (s)he had reached the end position. This measure was only calculated for the conditions in which there was haptic guidance, VS+H and VO+H, and the difference in this measure between the conditions was tested with a paired t-test.

The directions of the errors in the different conditions were assessed by comparing them to the directions of the visuo-haptic biases measured in condition VA. This was done by calculating, for each condition, the length of the projection of the error on the vector measured in condition VA. The ratio between the projected vector and the error vector was calculated to assess the amount of alignment between the two. Thus, a ratio of 1 means that the two vectors were oriented in exactly the same direction, a value of 0 means that they were oriented perpendicularly, and value of -1 means that the vectors were oriented in exactly opposite directions. The reason for calculating these values was to assess if the errors measured in the different conditions were related to the visuo-haptic biases. We expected that to be the case for VO+H, but finding a similar relation for VS+H would mean that the adaptation procedure in VA had not been completely successful.

Prior to the analyses, an outlier analysis was performed. For each target, participant, and condition, end points that were more than 4 standard deviations away from the mean of the data, when the mean was calculated without that particular end point, were considered to be outliers. This resulted in a total of 29 outliers, which was 0.4% of the data. Some of the outliers were caused by a mismatch between the two measurement systems, when participants accidentally had not pressed both space bars to proceed to the next trial. These mismatches were usually noticed quickly by the participants, since the guidance force was directed towards the wrong target if a mismatch occurred. For one participant, the visual image presented in the first condition was shifted in such a way that it was not visible any more for two of the targets, because it had crossed the limits of the monitors. For that participant, the data of those two targets were not used in the analysis.

8.3   Results

The accuracy of the responses is shown in Fig. 8.5a. For condition VA, in which the visual positions were adjusted according to the participant's responses, the mean difference between the target position and the positions of the visual image in the last 10 trials of each target position indicates the size of the visuo-haptic bias, which was 9.3 ± 1.9 cm (mean $\pm$ standard error). The visuo-haptic biases were mostly in the depth direction, as can also be seen in the typical example in Fig. 8.4. The size of the bias and the presence of components other than the one in the depth direction were different between participants, as can bee seen in Fig. 8.6. For the conditions other than VA, the distance between the mean target locations indicated by the participants and the target locations were 3.3 ± 0.6 cm for VS, 3.1 ± 0.6 cm for VS+H, and 5.8 ± 0.9 cm for VO+H. The repeated measures ANOVA showed a significant main effect of condition ($F_{1,2,12} = 19$, $p < 0.001$). Bonferroni-corrected post-hoc comparisons showed that VO+H differed significantly from both VS ($p < 0.0035$) and VS+H ($p < 0.0031$), while VS and VS+H did not differ significantly ($p = 1.0$). In conditions VS and VS+H, in which the shifted visual image was presented, the accuracy was the highest.

The precision of the responses is shown in Fig. 8.5b. The volumes of the 95% confidence ellipsoids were 284 ± 58 cm^3 for VS, 135 ± 28 cm^3 for VS+H,

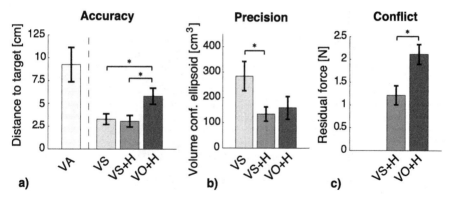

Fig. 8.5 Accuracy, precision and amount of conflict. All errors bars show mean ± one standard error. * Significant difference in post-hoc comparisons. (**a**) Accuracy of responses: the distance between indicated and physical target position. Data of the last 10 trials for each target in condition VA (adjusting visual information) are presented as a reference, showing the difference between the physical target position and the position of the visual image. Note that presentation of a shifted visual image (VS and VS+H) yields the highest accuracy. (**b**) Precision of responses: the volume of the 95% confidence ellipsoid. Adding haptic guidance increases the precision of the responses (VS+H and VO+H). (**c**) Conflict: the residual force that was still present on the handle when the participants indicated they had reached the target. The level of conflict is lower for the shifted visual image

Fig. 8.6 Magnitude of the visuo-haptic biases per participant, as measured in condition VA. Each bar represents one participant, and bars are ordered according to bias magnitude. Note that there is a wide range of biases across participants

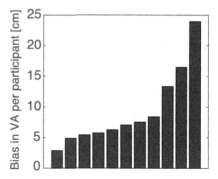

and $159 \pm 44 \, \text{cm}^3$ for VO+H. The repeated measures ANOVA again showed a significant main effect of condition ($F_{2,20} = 7.5$, $p = 0.0038$). Bonferroni-corrected post-hoc comparisons showed that only conditions VS and VS+H differed significantly ($p = 0.030$), while VO+H did not differ significantly from VS ($p = 0.086$) or VS+H ($p = 0.93$). Adding haptic guidance to the shifted visual image increased precision, which can also be seen in the typical example in Fig. 8.4.

The level of conflict between the haptic guidance and the participant is shown in Fig. 8.5c. It was $1.2 \pm 0.2 \, \text{N}$ for VS+H and $2.1 \pm 0.2 \, \text{N}$ for VO+H. A paired *t*-test showed a significant difference between these two conditions ($t_{10} = -5.2$, $p < 0.001$). The level of conflict was lower in the condition in which the visual image was shifted than in the condition with veridical visual information.

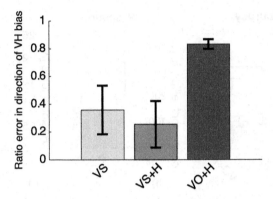

Fig. 8.7 Direction of the errors in conditions VS, VS+H and VO+H. The direction is expressed as a ratio of the projection of the error vector on the vector of the visuo-haptic bias, as measured in condition VA, and the length of the error vector itself. The error vector in condition VO+H is oriented in almost the same direction as the visuo-haptic bias

The assessment of the direction of the errors is shown in Fig. 8.7. The analysis yielded a ratio of 0.36 ± 0.18 for VS, 0.25 ± 0.17 for VS+H and 0.83 ± 0.03 for VO+H. Thus, the errors in condition VO+H were most closely aligned with the visuo-haptic biases measured in condition VA.

8.4 Discussion

In this study, we show that adjusting for user-specific biases in haptic guidance in 3D can be useful. We used an error-based paradigm in the first condition (VA) to establish the visual positions that corresponded to the haptic representations of the target positions. In condition VS, presenting shifted visual information indeed resulted in a smaller distance between the movement end points and the target positions, and thus a better accuracy. In condition VS+H, the addition of haptic guidance towards the target locations did not further improve the accuracy (compared to condition VS), but did result in smaller confidence ellipsoids than without haptic guidance, and thus a better precision. Finally, in condition VO+H, the precision was comparable to condition VS+H, but in condition VO+H, the accuracy was poorer: the end points were further away from the targets. Together, this shows that adding haptic guidance improves precision, but correcting for visuo-haptic biases has the added benefit of also increasing accuracy. Consequently, the user-specific guidance resulted in much smaller levels of conflict between participant and guidance than did veridical information. These results are in line with the results of the previous study on user-specific adjustments of haptic guidance in 2D [11]. It is difficult to directly compare the two studies, since the biases measured in Kuling et al. were much smaller, because they were measured in a 2D-setup. When comparing the relative increase in accuracy, the previous study showed about a 20% increase in accuracy when individualizing guidance, while the current study shows about a 50% increase. However, this does not necessarily mean that one type of correcting is more useful than the other, since the setups and the absolute sizes of the biases were very different which could influence the usefulness of the individualization.

Even though the accuracy of the participants increased in VS and VS+H, the accuracy values do not have a mean of 0 for these conditions, while a perfect correction had yielded no bias in case of a sufficiently large sample size. Partly, these residual biases are caused by an imperfect adaptation in condition VA. The analysis following the experiments showed some outliers for condition VA. For the online calculation of the position of the shifted images to be presented in conditions VS and VS+H, which was based on the last 10 trials of condition VA, all the data were used. However, removing these outliers would have only slightly improved the accuracy values. Moreover, the previous study in 2D [11] also reported non-zero biases for the user-specific guidance conditions, while there was no adaptive phase present. The direction analysis can also shed some light on this problem. The results from condition VA showed that the participants' visuo-haptic biases tended to be mainly oriented in the depth direction. Participants usually overshot the target on the first trial, after which the visual image needed to be shifted closer to them in order for them to bring their hand to the target location. This overshooting behaviour is in accordance with the literature [e.g. 2, 19, 23]). One reason for finding decreased precision in the depth direction is purely geometrical. When assuming that the targets are judged from retinal separations or changes in eye orientation, errors in angular judgements have a much larger effect in the depth direction that in the vertical plane [3]. It has been shown that for targets with limited depth information, presented in the dark, participants usually show an underestimation for the distance of far targets and an overestimation for near targets [19]. Relying on vergence cues results in increasing undershooting behaviour with increasing target distance [13]. None of our targets were presented very close to the body, and the undershooting behaviour was most prominent for the furthest targets. The direction analysis also showed that participants made errors that were close to the direction of the individual visuo-haptic biases in the condition with veridical information (VO+H). This is not surprising, since we corrected for the biases in these conditions. In the two conditions using shifted visual information, however, the directions of the errors were not aligned with the visuo-haptic biases. This again points to the residual errors in condition VS and VS+H not being directly related to the visuo-haptic biases, showing that they are not caused by an imperfect adaptation in condition VA. Thus, the origin of these residual biases remains unclear.

In a practical teleoperation situation, a shift of the visual image should be easy to implement if the visual information is provided through a screen anyway. Since the visuo-haptic biases are stable over the time span of at least one month [12], it would be sufficient to measure the biases once a month (and possibly even less often) in a simple pointing task. The biases could then be used as part of an operator-specific setup of the devices, which would determine the operator-specific shift of the visual information. An example of a system in which such an adjustment could be useful is a virtual reality based system in which hand tracking is not present, so the user's hands are not displayed. A potential problem of adjusting for visuo-haptic biases is that, to capture the biases throughout the workspace, either a lot of points throughout the workspace or a better understanding of the mechanism of the biases would be needed. If future research would provide us with a model of these biases, it would be far easier to make the proper adjustment throughout the workspace or even predict biases by modelling them for the specific user.

This study shows that individualizing guidance results in a higher accuracy and a lower level of conflict between operator and guidance than traditional, veridical guidance does. So, perceptually consistent haptic guidance seems to have benefits over physically correct guidance. In addition to these immediate benefits, we think that the added benefit of reducing the amount of conflict is that this probably increases the operator's trust in the system, since the system is more intuitive to use. It has been shown that intuitiveness is a very important factor in the usefulness of haptic guidance [14]. When the same amount of information is present, a more intuitive type of haptic guidance results in a better operator performance. This means that user-specific adjustments to haptic guidance show potential to increase operator performance, decrease the amount of conflict between operator and machine and make the contact between operator and machine more intuitive.

8.5   Conclusions

In this study, we have investigated the usefulness of correcting for user-specific biases in human perception in the design of haptic guidance. As an example of user-specific biases, we used the paradigm of visuo-haptic biases. Our data show that user-specific guidance increased accuracy and decreased the amount of conflict between operator and machine. The precision was improved in all conditions in which haptic guidance was presented. These results show the potential of correcting for user-specific perceptual biases when designing haptic guidance.

References

1. Abbink DA, Mulder M, Boer ER (2012) Haptic shared control: smoothly shifting control authority? Cogn Tech Work 14(1):19–28
2. Adamovich SV, Berkinblit MB, Fookson O, Poizner H (1998) Pointing in 3D space to remembered targets. I. Kinesthetic versus visual target presentation. J Neuropsychol 79(6):2833–2846
3. Brenner E, Smeets JB (2000) Comparing extra-retinal information about distance and direction. Vis Res 40(13):1649–1651
4. Fuentes CT, Bastian AJ (2010) Where is your arm? Variations in proprioception across space and tasks. J Neurophysiol 103(1):164–171
5. Gentaz E, Hatwell Y (2004) Geometrical haptic illusions: the role of exploration in the Müller-Lyer, vertical-horizontal, and Delboeuf illusions. Psychon Bull Rev 11(1):31–40

6. Haggard P, Newan C, Blundell J, Andrew H (2000) The perceived position of the hand in space. Percept Psychophys 62(2):363–377
7. Hayward V, Astley OR, Cruz-Hernandez M, Grant D, Robles-De-La-Torre G (2004) Haptic interfaces and devices. Sens Rev 24(1):16–29
8. Helms Tillery SI, Flanders M, Soechting JF (1994) Errors in kinesthetic transformations for hand apposition. Neuroreport 6(1):177–181
9. Kuling IA, Brenner E, Smeets JBJ (2013) Proprioception is robust under external forces. PLoS ONE 8(9):e74,236
10. Kuling IA, Brenner E, Smeets JBJ (2015) Torques do not influence proprioceptive localization of the hand. Exp Brain Res 233(1):61–68
11. Kuling IA, van Beek FE, Mugge W, Smeets JBJ (2016) Adjusting haptic guidance to idiosyncratic visuo-haptic matching errors improves performance in reaching. IEEE Trans Hum-Mach Syst 46:921–925
12. Kuling IA, Brenner E, Smeets JBJ (2016) Errors in visuo-haptic and haptic-haptic location matching remain consistent over long periods of time. Acta Psychol 166:31–36
13. Mon-Williams M, Tresilian JR (1999) Some recent studies on the extraretinal contribution to distance perception. Perception 28(2):167–181
14. Mugge W, Kuling I, Brenner E, Smeets J (2016) Haptic guidance needs to be intuitive not just informative to improve human motor accuracy. PLoS ONE 11(3):e0150912
15. Pierce R, Kuchenbecker K (2012) A data-driven method for determining natural human-robot motion mappings in teleoperation. In: 4th IEEE RAS EMBS international conference on biomedical robotics and biomechatronics (BioRob), pp 169–176
16. Rincon-Gonzalez L, Buneo CA, Helms Tillery SI (2011) The proprioceptive map of the arm is systematic and stable, but idiosyncratic. PLoS ONE 6(11):e25214
17. Smeets JBJ, van den Dobbelsteen JJ, de Grave DDJ, van Beers RJ, Brenner E (2006) Sensory integration does not lead to sensory calibration. Proc Natl Acad Sci 103(49):18781–18786
18. Soechting JF, Flanders M (1989) Sensorimotor representations for pointing to targets in three-dimensional space. J Neurophysiol 62(2):582–594
19. Sousa R, Brenner E, Smeets JBJ (2010) A new binocular cue for absolute distance: disparity relative to the most distant structure. Vis Res 50(18):1786–1792
20. Sousa R, Smeets JBJ, Brenner E (2012) The effect of variability in other objects' sizes on the extent to which people rely on retinal image size as a cue for judging distance. J Vis 12(6):1–8
21. Van Beek FE, Bergmann Tiest WM, Gabrielse FL, Lagerberg BWJ, Verhoogt TK, Wolfs BGA, Kappers AML (2014) Subject-specific distortions in haptic perception of force direction. In: Auvray M, Duriez C (eds) Haptics: neuroscience, devices, modeling, and applications, Part I, vol 8618. Lecture notes in computer science. Springer, Berlin/Heidelberg, pp 48–54
22. Van Beers RJ (2009) Motor learning is optimally tuned to the properties of motor noise. Neuron 63(3):406–417
23. Van Beers RJ, Sittig AC, Denier van der Gon JJ (1998) The precision of proprioceptive position sense. Exp Brain Res 122(4):367–377
24. Van der Kooij K, Brenner E, van Beers RJ, Schot WD, Smeets JBJ (2013) Alignment to natural and imposed mismatches between the senses. J Neurophysiol 109(7):1890–1899
25. Van der Linde R, Lammertse P (2003) HapticMaster – a generic force controlled robot for human interaction. Int J Ind Robot 30(6):515–524

Chapter 9
General Discussion

Abstract The aim of this book was to investigate parameters in haptic perception that are important for designing haptic devices and haptic feedback. The data gathered in this investigation cover fundamentals of static haptic perception, fundamentals of dynamic haptic perception and studies into the design of intuitive haptic guidance. In this discussion chapter, these data will be discussed in the light of the aim of the book. First, the fundamental insights which this book provides for haptic perception will be discussed. Next, the reproducibility of the results is discussed. Finally, recommendations on how these results could be applied to haptic device and haptic guidance design will be given.

9.1 Fundamental Insights

As stated in the Introduction (Chap. 1), biases and discrimination thresholds are the main parameters that are used in psychophysical research to describe human performance [13]. Knowing these parameters can help to determine how to design your haptic device in a way that it fits the human needs and is intuitive to use. Using this fundamental approach can be seen as working from general knowledge (knowledge on human perception) to a more specific application (the design of a particular device), so this approach could be called a deductive one [14]. In most technology-oriented research, the problem is usually approached from the other direction: devices are designed and tuned based on previous designs and tunings, and then they are tested in a human factors experiment. If the results are satisfactory, the device is deemed good enough. If not, the tuning or design is changed and the human factors experiment is repeated. When a lot of these experiments have been performed, general design rules might be inferred from the experiments, so this would be a more inductive way of building theory [9]. Obviously, there are advantages to the inductive approach: designers perform tests on the parameter that they are actually interested in, which is the performance of the operator when using that particular device. However, the pitfall of this approach is that it is difficult to understand why some devices and tuning parameters perform well and others do not. When no fundamental knowledge on the problem is gained, the best-case-

© Springer International Publishing AG 2017

F.E. van Beek, *Making Sense of Haptics*, Springer Series on Touch and Haptic Systems, https://doi.org/10.1007/978-3-319-69920-2_9

scenario is that general guidelines can be formulated. So, the approach in this book was a deductive one: by investigating human perception, information about the precision and accuracy of perception was obtained. This information can be used to understand how to design haptic devices in general, since the information does not apply to one specific device.

9.1.1 Part I: Static Perception

The first part of the book describes a situation in which haptic information is acquired in a static fashion, so in which stimuli are applied to the stationary human hand. The obvious parameter to look at in this context is force perception: how do humans perceive that someone is pulling or pushing their hand? In literature, a lot of attention has been devoted to investigating *discrimination thresholds* of force perception (see Jones [12] for a review). However, *biases* in force perception in directions other than that of gravity have hardly been studied, so that is the topic of the first part of this book.

In Chaps. 2 and 4, direction-dependent biases in force magnitude perception were found, which were very consistent across participants. Forces exerted perpendicular to the line between shoulder and hand were perceived as being 50% larger than forces exerted along this line. These biases were not directly related to the measured arm dynamics parameters, so the arm dynamics parameters cannot be the sole contributors to the perceptual effect, but they can still play a role in causing it.

In Chaps. 2 and 3, direction-dependent biases were also found in force direction perception. The biases ranged between −30° to 60° and varied greatly between participants and force directions, resulting in participant-dependent error patterns. No groups of similar patterns could be discerned within the total group of participants, and the patterns could not be explained using general subject parameters, like arm length (Chap. 3). Nonetheless, the patterns were found to be consistent within participants, even when they were measured a month later (Chap. 3).

9.1.2 Part II: Dynamic Perception

The second part investigates what happens when humans start moving their hands. Firstly, they need to sense the hand movements, so the perception of movement is the first topic of interest. Secondly, they make hand movements in order to interact with something, so the perception of objects during movement is the second topic of interest.

In Chap. 5, discrimination thresholds for movement distance were investigated using a task in which two movement distances had to be compared. In contrast to the literature on force perception, the literature on movement distance mainly reports *biases* [e.g. 5, 7, 10, 34] and has not devoted much attention to *discrimination thresholds*. So, in this study, discrimination thresholds were tested in conditions that were known to affect biases in movement distance. In most conditions, we found no effect of that condition on the discrimination threshold. Generally speaking, for movement distances of 25 and 35 cm, a Weber fraction of about 11% was found along all cardinal axes. For passive movements, the threshold was a bit higher, and adding cutaneous information did not improve the precision. Movement parameters, which were recorded during the task, showed that participants generally adopted the strategy to keep the speed profiles of two movements the same and to compare the resulting movement times in order to infer the movement distances.

In Chap. 6, the perception of object hardness was investigated for different types of movements. Object hardness is an important object property in teleoperation tasks, since a proper perception of it is essential to assess how much force is needed to manipulate the object, while not breaking it. A large effect of damping on perceived hardness was found. This effect was task-dependent: for an in-contact task (in which the movement started at the object's surface), adding global damping increased perceived hardness, while for a contact-transition task (in which there was a free-air phase before making contact), adding global damping decreased perceived hardness. The latter effect was much larger than the former. The movement parameters revealed that this task-dependency was not caused by a change of movement strategy, since for both tasks in all experiments, participants used the same parameters to base their perceptual decision on. So, an actual difference in task dynamics probably caused the task-dependency of the effect. In both Chaps. 5 and 6, analyzing the movement parameters helped to understand the strategies responsible for the perceptual outcomes. In perceptual research, movement parameters are not always investigated, not even in dynamic tasks. The results in this book suggest that it would be interesting to include an analysis of movement parameters in future research on dynamic tasks.

9.1.3 Part III: Applications

The results in the last part of this book, consisting of Chaps. 7 and 8, could be applied more directly to haptic guidance. However, the chapters also provide insight in multisensory integration, since they involve the perception of multiple sources of information.

Chapter 7 revolves around the integration between position and force information in a task of finding the center of a weak force field. Of course, this easily relates to force fields used in haptic guidance, but it also provides fundamental information about the integration of information. It has been shown that humans can integrate information from multiple information sources [e.g. 33]. The data in this chapter show that in this task, force and position are not integrated into a percept of stiffness, but humans rather simply use the position(s) where the force reaches the detection threshold level to estimate the center of the force field. This demonstrates that information does not have to be integrated, even if multiple sources of information are available. The lack of integration could be caused by the serial nature of the task: obtaining enough information to make a stiffness estimate would require storing information from serial exploration, which is a costly process [6, 18], while the observed strategy is of a far less serial nature and might therefore be preferred.

Chapter 8 is most interesting from an application point of view, since it mainly shows that correcting for perceptual biases in haptic guidance can increase user performance. In addition, it shows that visuo-haptic biases in 3D are mostly oriented along the depth direction. Most participants overshot the target, when pointing to a visual target without seeing their hand. This is in accordance with literature on similar tasks [1, 29, 32].

In almost all the chapters, significant perceptual biases were reported. It is thus safe to say that human perception is often not veridical, which means that the perceptual world often does not scale linearly with the physical measures that we use to describe the world with. So, why are we still able to perform actions at all? Why don't we bump in to objects all the time, as, for instance, the perceived length of our arm movements depends on our movement direction? It could be that we have learned strategies to cope with these biases, since we have encountered them throughout our lives. Interestingly, humans tend to believe that others are prone to be biased, but usually they have the feeling that their own observations are bias-free, a phenomenon called 'the bias blind spot' [25, 26]. Of course, humans might be able to correct for their biases subconsciously, even when they are not aware of their own biases. However, it would not make much sense that human beings, who have evolved over the course of millions of years, would end up with being stuck with magically imposed biases for which they have to correct in all their actions. It is far more probable that these biases actually reveal functional systems in sensorimotor control. In other words: perceptual biases are probably often related to the action connected to the percept. In Chap. 4, we could not find a direct link between force magnitude perception and arm biomechanics, but links between perception and action have been described throughout the perceptual literature [e.g. 2, 19]. So, in future research, it would be insightful to not only measure perceptual biases, but to also speculate—and if possible perform experiments—on the link between perceptual biases and the action connected to the perceived property.

9.2 Reproducibility

An important question in all fields of research is the reproducibility of results. Recently, psychological research received a particularly hard blow when a large consortium tried to reproduce 100 psychological studies, which had all been published in high-quality journals [22]. Whereas in the original publications 97 of the 100 papers reported a significant effect, only 36 of the 100 replications showed a significant effect. Fortunately, hardly any of the results in this book rely on a single set of data, as will be shown in this section.

In the two chapters on force magnitude perception, Chaps. 2 and 4, very similar biases were found, even though the former chapter describes biases measured in 2D using a custom-made setup, while the latter chapter describes biases measured in 3D using the HapticMaster (compare Figs. 2.4 and 4.4). The data on arm biomechanics in Chap. 4 have not been reproduced in this book, but are very comparable to values reported in literature [15, 24]. For the two chapters on perception of force direction, Chaps. 2 and 3, it is a bit harder to draw conclusions, since there were larger differences between participants and different participants took part in the different experiments. However, what is apparent is that most participants showed considerable perceptual biases in at least some directions, and all data sets show large variations between participants (compare Figs. 2.6a, 3.2, and 3.3). The conclusions in Chap. 3 on the origin and consistency of the error patterns are the only conclusions in this part of the book that rely on a single set of data. The absence of a correlation between the participant's general characteristics and the error patterns in force direction perception could be due to a group size ($n = 25$) that might be too small for this type of correlational research. However, none of the p-values were close to significant (all $p \geq 0.15$), so it is not very likely that a larger group size would have yielded significant results. The other experiment in this chapter, showing the consistency of the patterns within participants, did not yield reproducible patterns for all participants. However, for most participants the patterns were reproducible, therefore the conclusion was drawn that generally speaking, the patterns are reproducible. A replication of this experiment using a larger group of participants could be useful in order to draw firmer conclusions.

The two chapters on perception of dynamic parameters also show consistent results. The data on discrimination of movement distance, presented in Chap. 5, have not been reproduced in this book, but all the different research questions yielded biases in the same magnitude range (see Fig. 5.4). For research questions in which a main effect of condition was found, this effect was highly significant (all $p \leq 0.003$). The data on the influence of damping on hardness perception, Chap. 6, were reproduced in the same experiment, since Experiments 1 and 2 were performed with different groups of participants and yielded comparable results (compare Figs. 6.5 and 6.6). Experiment 3 had to be repeated with a different group of participants because of errors in the measuring procedure (only the data of the last group are shown in Chap. 6), but the results of the replications were very similar.

Finally, the two more applied chapters, Chaps. 7 and 8, were both follow-up studies on themes that some of the authors had been working on previously. The current results are all in line with the results from the previous studies. The pattern in the biases found in the chapter on the integration of force and position information are very comparable to the biases reported in Baud-Bovy [4]. The results of the chapter on adjusting haptic guidance to individual biases are also very comparable to the results in Kuling et al. [16], even though the biases and effect sizes in the previous study in 2D were much smaller than those in the current study in 3D.

Summarizing, most of the data in this book have been reproduced in similar experiments in one of the chapters or are comparable to data in similar experiments in literature, which ensures that the chance that the reported effects are reproducible is considerable. It is hard to make claims on the generalizability of the results, since it is known in perceptual research that small adjustments to research designs can have massive effects on the size of perceptual biases (see for instance the review by McFarland and Soechting [20] on all the factors influencing the size of the radial-tangental illusion). Some of the experiments in this book have been reproduced using different devices, such as the force magnitude perception experiment, which has first been performed on a custom-made device (Chap. 2) and then reproduced using a HapticMaster (Chap. 4), while yielding the same results. The experiment on correcting for visuo-haptic biases (Chap. 8) was performed on a HapticMaster, while the first experiment in that research line was performed on a PHANToM Premium [16]. However, these are all still 'devices', so whether these results can be translated to a natural situation in which you perform actions with your own hands, without any restrictions on, for instance, hand movement, remains to be investigated.

9.3   Implications for Design

The results from the more fundamental parts of this book also provide information on human perception that can be useful to consider when designing haptic devices. In the next sections, the possible applications of this fundamental knowledge are explained.

9.3.1   *Haptic Parameters for Haptic Device Design*

Chapters 2 and 3 show that perception of force direction is not veridical, and it is also very different between participants. A general observation that can be drawn from these results is that it is probably not very useful to use force direction as a way to communicate information. Mean individual biases ranged between $-30°$ and $60°$, so one physical force direction can result in a perceptual direction with a range of $90°$. Note that this does not mean that differences in direction cannot be perceived reliably; the *discrimination threshold* for force direction is $30°$ [3, 11, 30], so differences in force direction exceeding this value can be perceived reliably. However, this book shows that the *accuracy* of the perception of force direction is generally very poor.

It is probably not necessary to build haptic devices that outperform the human capacities, so the results on the precision of distance perception (Chap. 5) show that, to communicate distance information, a precision better than 11% of the movement distance is acceptable. For the smallest movement distance in our study (15 cm), the Weber fraction was higher than for the other two distances, so probably there is an absolute limit in precision for smaller distances, but it is likely that we did not reach that yet at a movement distance of 15 cm. The data also show that adding cutaneous information does not improve precision, so in order to convey precise distance information, it is not necessary to simulate surfaces.

The results on the influence of damping on perceived hardness (Chap. 6) show that adding global damping to a system has a large effect on contact-transition tasks. Since contact-transition situations are common in teleoperation applications [27], designers should be careful about adding damping to their system to increase its stability, if the operator is required to have a proper perception of object hardness. As a start of a guideline, Fig. 6.7 could be used, which describes the interplay between stiffness and damping in perceived hardness. Of course, these values have been measured on a specific device (the HapticMaster) in particular tasks, so there is no guarantee that the values are the same for other tasks or devices. Therefore, these specific values can only be seen as a start of a more general guideline. In general, the best solution from a human perspective would be to avoid inserting (a lot of) damping at all, which is possible when using less conventional designs, such as the one proposed in Heck, Saccon and Nijmeijer [8]. These authors propose a scheme in which the energy flows of the master and the slave are constantly monitored, and damping is only inserted when the system's stability is in danger.

9.3.2 Haptic Parameters for Haptic Guidance Design

How to design the force field in haptic guidance in a way that is intuitive for the human operator? This book surely does not provide a full answer to this question, but the presented data do provide ideas on adjustments that could make haptic guidance more intuitive.

Chapters 2 and 4 show that perception of force magnitude is not veridical, but shows a direction-dependent distortion. This means that it might be useful to correct for this distortion when the aim is to provide the same message throughout the force field, independent of the direction of the force. So, it might be useful to design a force field that is perceptually equal in all directions, rather than physically equal. It is not clear how to do this yet, since the data in these chapters have been measured using one arm posture, while in a real teleoperation situation, the operator probably moves his arm. Unfortunately, no direct relationship between arm dynamics and force magnitude perception was found in Chap. 4. There is a strong relation between arm dynamics and posture [21, 31], so if arm dynamics had been directly related to the perceptual distortion, arm posture could have been used to predict the effects on the perceptual distortion. In future research, it would therefore be useful to measure the distortion in force magnitude perception for various arm postures. This could shed light on relations between the perceptual biases and arm posture, and possibly also on their relation with arm dynamics.

The design of force fields for haptic guidance that are intuitive to use is not very straightforward, as shown in Chap. 7. This chapter shows that, when trying to find the center of a force field, which is also the basic task in a haptic guidance situation, humans do not integrate position and force information into a stiffness percept. Our study showed some consequences of this notion, relating to asymmetric force fields, uni-lateral force fields and visual information. In asymmetric force fields, large biases in finding the center of the force field were found, which were oriented towards the weaker side of the field. A similar result was found for uni-lateral force fields, in which participants were asked to find the edge of the force field. In that situation, participants always perceived the edge of the force field to be somewhat inside the force field. The latter finding raises an interesting point for haptic guidance applications, as operators usually do not overshoot the target very far. Therefore, the chance that their movement reaches the part of the force field with forces above the detection threshold level behind the target is not very large, thus operators are mainly exposed to a uni-lateral situation. In this situation, participants perceive the center of the force field, and thus the target, to be at the position where the force reaches the detection threshold level. So, it could be worthwhile to position the force field in such a way that the force magnitude at the target is not 0, but is at the detection threshold level. This does make the design of the force field more complicated when the force field needs to be multi-dimensional, but it would be worthwhile to test this approach in future research. A final observation from this study is that visual information about hand position, while it was not informative about the position of the force field, still caused an increase in precision. So, depending on which types of information the operator uses to perform a task, it might be worthwhile to provide visual information, even if this seems to be pointless for the direct task objective. All the observations in this chapter could be explained using a mathematical model. When trying to assess the perceptual consequences of haptic guidance design choices, this model could be used to obtain a general idea of the direction of the perceptual effects.

In this book, a lot of studies report biases in human perception, of which some are participant-specific, such as the biases in the perception of force direction (Chaps. 2 and 3). It has already been shown that correcting the mapping between operator and slave movements, by using parameters that are consistent across participants, increases user performance [23]. In the final chapter, Chap. 8, we aimed to test if it is also useful to correct for participant-specific biases in haptic perception. To investigate this, we wanted to use a well-known paradigm for participant-specific biases, so we chose the paradigm of visuo-haptic biases [28]. The task for the participants was to move their unseen hand to a visual target, which usually results in large errors. First, the biases were measured, after which two types of haptic guidance were compared: one that was adjusted to correct for the biases and one that was not. Adjusting for the biases significantly improved user performance. In our study, the feedback was adjusted by correcting the visual information: the visual image was shifted, while the haptic guidance was directed towards the original target position. In the first experiment on this idea, a similar approach was used, but in that case, the position of the haptic guidance was corrected, while the visual image remained at the target location [16]. In that approach, participants did not end up at the target location, but at a location shifted by their visuo-haptic bias. However, since the bias was known, their end position was predictable. This was the objective of the study, since in a teleoperation setting, predictable biases on the master side can be corrected for on the slave side, which ultimately ensures that the movement of the slave reaches the target position. Nonetheless, our approach in Chap. 8 seems to be more practical: the only adjustment that is needed to increase operator performance is a shift of visual information. It is noteworthy that the measurement of the visuo-haptic bias was a bit more tricky in our approach: usually the visuo-haptic bias congruent with a visual position is measured, which can easily be done by asking participants to point to a visual target, as was also done in Kuling et al. [16]. In the current approach, we needed to measure the bias congruent with a haptic location, so an adaptive paradigm was used to find the visual location which matched the haptic representation of the target location. In a practical setting, it is imaginable to measure the operator-specific biases once every month using a task that is comparable to our adaptive paradigm, after which the personalized settings can be used throughout that month. Most of these biases are fairly stable over time (Chap. 3 and Kuling et al. [17]), so the benefits should be fairly long-lasting.

In this book, it has been shown that human perception is often not veridical, but shows consistent and reproducible biases in both static and dynamic perception. This information can be used in the design of haptic feedback devices and in the design of haptic guidance. Taking human perception into consideration when building devices that humans need to operate is a logical step to enable the construction of intuitive, human-centered systems.

References

1. Adamovich SV, Berkinblit MB, Fookson O, Poizner H (1998) Pointing in 3D space to remembered targets. I. Kinesthetic versus visual target presentation. J Neurophysiol 79(6):2833–2846
2. Ahmed AA, Wolpert DM, Flanagan JR (2008) Flexible representations of dynamics are used in object manipulation. Curr Biol 18(10):763–768. https://doi.org/10.1016/j.cub.2008.04.061
3. Barbagli F, Salisbury K, Ho C, Spence C, Tan HZ (2006) Haptic discrimination of force direction and the influence of visual information. ACM Trans Appl Percept 3(2):125–135
4. Baud-Bovy G (2014) The perception of the centre of elastic force fields: a model of integration of the force and position signals, chap 7. Springer series on touch and haptic systems. Springer, London, pp 127–146
5. Bergmann Tiest WM, van der Hoff LMA, Kappers AML (2011) Cutaneous and kinaesthetic perception of traversed distance. In: Proceedings of the IEEE world haptics conference, pp 593–597
6. Dopjans L, Bülthoff HH, Wallraven C (2012) Serial exploration of faces: comparing vision and touch. J Vis 12(1–6):1–14
7. Faineteau H, Gentaz E, Viviani P (2003) The kinaesthetic perception of Euclidean distance: a study of the detour effect. Exp Brain Res 152(2):166–172
8. Heck D, Saccon A, Nijmeijer H (submitted) Direct force-reflecting two-layer approach for stable bilateral teleoperation with time delays
9. Heit E (2000) Properties of inductive reasoning. Psychon Bull Rev 7(4):569–592
10. Hermelin B, O'Connor N (1975) Location and distance estimates by blind and sighted children. Q J Exp Psychol 27(2):295–301
11. Ho C, Tan HZ, Barbagli F, Salisbury K, Spence C (2006) Isotropy and visual modulation of haptic force direction discrimination on the human finger. In: Proceedings of Eurohaptics 2006, pp 483–486

12. Jones LA (1986) Perception of force and weight: theory and research. Psychol Bull 100(1): 29–42
13. Jones LA, Tan HZ (2013) Application of psychophysical techniques to haptic research. IEEE Trans Haptic 6(3):268–284
14. Kimmig A (2013) Deductive reasoning. Springer, New York, pp 557–558
15. Krutky MA, Trumbower RD, Perreault EJ (2009) Effects of environmental instabilities on endpoint stiffness during the maintenance of human arm posture. In: Annual international conference of the IEEE engineering in medicine and biology society, pp 5938–5941
16. Kuling IA, van Beek FE, Mugge W, Smeets JBJ (2016) Adjusting haptic guidance to idiosyncratic visuo-haptic matching errors improves performance in reaching. IEEE Trans Hum-Mach Syst 46:921–925
17. Kuling IA, Brenner E, Smeets JBJ (2016) Errors in visuo-haptic and haptic-haptic location matching remain consistent over long periods of time. Acta Psychol 166:31–36
18. Loomis JM, Klatzky RL, Lederman SJ (1991) Similarity of tactual and visual picture recognition with limited field of view. Perception 20(2):167–177
19. McCloskey D, Ebeling P, Goodwin G (1974) Estimation of weights and tensions and apparent involvement of a 'sense of effort'. Exp Neurol 42(1):220–232
20. McFarland J, Soechting JF (2007) Factors influencing the radial-tangential illusion in haptic perception. Exp Brain Res 178(2):216–227
21. Milner TE (2002) Contribution of geometry and joint stiffness to mechanical stability of the human arm. Exp Brain Res 143(4):515–519
22. Open Science Collaboration (2015) Estimating the reproducibility of psychological science. Science 349(6251)
23. Pierce R, Kuchenbecker K (2012) A data-driven method for determining natural human-robot motion mappings in teleoperation. In: 4th IEEE RAS EMBS international conference on biomedical robotics and biomechatronics (BioRob), pp 169–176
24. Pierre M, Kirsch R (2002) Measuring dynamic characteristics of the human arm in three dimensional space. In: Proceedings of the 24th annual conference and the annual fall meeting of the biomedical engineering society EMBS/BMES conference, engineering in medicine and biology, vol 3, pp 2558–2560
25. Pronin E (2007) Perception and misperception of bias in human judgment. Trends Cogn Sci 11(1):37–43
26. Pronin E, Lin DY, Ross L (2002) The bias blind spot: perceptions of bias in self versus others. Personal Soc Psychol Bull 28(3):369–381
27. Sarkar N, Yun X (1996) Design of a continuous controller for contact transition task based on impulsive constraint analysis. In: Proceedings of the 1996 IEEE international conference on robotics and automation, vol 3, pp 2000–2005
28. Soechting JF, Flanders M (1989) Sensorimotor representations for pointing to targets in three-dimensional space. J Neurophysiol 62(2):582–594
29. Sousa R, Brenner E, Smeets JBJ (2010) A new binocular cue for absolute distance: disparity relative to the most distant structure. Vis Res 50(18):1786–1792
30. Tan HZ, Barbagli F, Salisbury K, Ho C, Spence C (2006) Force-direction discrimination is not influenced by reference force direction. Haptics-e 4:1–6
31. Tsuji T, Morasso P, Goto K, Ito K (1995) Human hand impedance characteristics during maintained posture. Biol Cybern 72(6):475–485
32. Van Beers RJ, Sittig AC, Denier van der Gon JJ (1998) The precision of proprioceptive position sense. Exp Brain Res 122(4):367–377
33. Van Beers RJ, Sittig AC, Denier van der Gon JJ (1999) Integration of proprioceptive and visual position-information: an experimentally supported model. J Neurophysiol 81(3):1355–1364
34. Wong TS (1977) Dynamic properties of radial and tangential movements as determinants of the haptic horizontal-vertical illusion with an L figure. J Exp Psychol Hum Percept Perform 3(1):151–164

Printed in the United States
By Bookmasters